SATELLITE IMAGERY FOR THE MASSES

f[illegible] [illegible]on

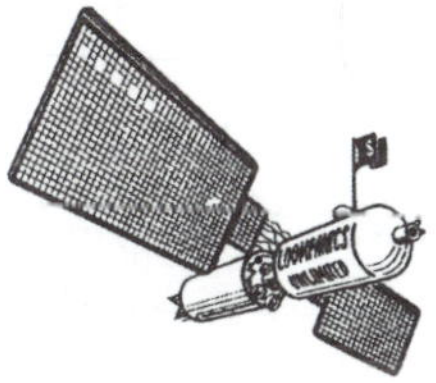

Loompanics Unlimited
Port Townsend, Washington

Neither the author nor the publisher assumes any responsibility for the use or misuse of information contained in this book. It is sold for informational purposes only. Be Warned!

Satellite Imagery for The Masses
How to Use and Profit from the Satellite Revolution

Cover by Harlan Kramer

Published by:
Loompanics Unlimited
PO Box 1197
Port Townsend, WA 98368
Loompanics Unlimited is a division of Loompanics Enterprises, Inc.
Phone: 360-385-2230
Fax: 360-385-7785
E-mail: service@loompanics.com
Web site: www.loompanics.com

ISBN 1-55950-240-1
Library of Congress Card Catalog Number 02003115256

Contents

Chapter One
Introduction....................1

Chapter Two
A Brief History of Satellite Imagery....................11

Chapter Three
How Satellite Imagery Works....................25

Chapter Four
What Satellite Imagery Can Do....................45

Chapter Five
A Thumbnail Guide to Interpreting....................59

Chapter Six
How to Make Money With Satellite Imagery....................115

Chapter Seven
Acquiring Satellite Imagery....................131

Chapter Eight
Satellite Imagery and Your Privacy....................143

Chapter Nine
Satellite Imagery and Three Dimensional Visualizations....................159

Appendix I
Picking the Right System for Your Needs....................167

Appendix II
Sources for Satellite Imagery and Other Resources 169

Appendix III
America's First Spy Satellite Mission — 9009 175

Glossary 187

To Nancy

Without whose love, support and persistence, this book would have not been written.

Acknowledgements

I couldn't have written this book without a lot of support. My greatest thanks goes to Sandy Perry and Luke. It was Sandy that got me interested in satellite imagery many years ago and was patient enough to answer all my questions — then and now. It was Luke who showed me how to take that imagery and make a living by finding weapons facilities.

I also want to acknowledge my friend and professional colleague, Ernie Ross, who unfortunately died just as I was beginning this book. It was Ernie who accepted my first article that used satellite imagery and continued to encourage me through the years. He will be remembered and missed.

A special thanks goes to Michael Patrick and the Christian Broadcasting Network News (CBN News) who were pioneers in the use of satellite imagery in investigative reporting. Then there are the satellite imaging companies, Space Imaging, SPOT, DigitalGlobe, Autometric, and Sovinformsputnik, who kindly allowed me to use their imagery in this book.

And a heartfelt thanks to the friends who gave me moral support and patiently listened to me babble endlessly about satellites while I wrote this — Rob, Dave, Lowell, Lavorris, Bill, Bob, Erwin, Melissa, and Marv. I'll try to talk about something else now.

And, finally, a thanks to Enzo, my cat and self-appointed editor, who sat in my lap the whole time I wrote this and made corrections as necessary.

Chapter One
Introduction

Imagine designing a computer game that shows exactly how it looked to American military pilots as they flew over Hanoi during the Vietnam War. Or getting paid by a news agency to find a top secret North Korean weapons facility. Imagine owning your own consulting business where you help other businessmen decide where they should place their new businesses. How about making a little money on the side finding a good placer gold deposit or antiques along an old pioneer trail? These are all possible today because of the availability of satellite imagery to the average person.

The last fifteen years have been to commercial satellite imagery what the 1980s and early 1990s were to computers. What was once technology limited to scientists is now found on the main streets of America. Everyday people like real estate agents, police, marketing experts, and journalists use satellite imagery on an everyday basis. In fact today, there are people using satellite imagery in their jobs that once thought

SPOT was something you found on a dress, not a satellite in space.

Today, satellite imagery is used in courts as legal evidence, by earth resources companies to find minerals, by environmentalists to find pollution, by historians to better understand history, and by civil engineers to build infrastructure. It is even being used by the government to tax you and find zoning violations.

There is no better example of how commercial satellite imagery has matured as an industry than what happened in the last two wars in Iraq.

In August 1990, the 82nd Airborne Division had a problem. They had been deployed to Saudi Arabia to stand between an Iraqi Army poised to attack and the Saudi oil fields. These few thousand lightly armed troops only had days in which to move their equipment across the treacherous sands of the Saudi Desert and into position where they could defend the oil fields against the fourth largest army in the world.

Without heavy weapons, the units could rely on only two factors to stop the Iraqi juggernaut: professional, trained soldiers and high tech equipment. But, moving soldiers and equipment was difficult because the soft desert sand could trap and disable the vehicles as effectively as an Iraqi minefield. At the same time, the available maps of Saudi Arabia were more than a quarter century old. If the allied forces were to confront the Iraqis, they would need accurate maps immediately.

Because traditional map making took months, the Army used commercial satellite imagery and the Global Positioning System (GPS). In addition, when air operations started, military pilots used commercial satellite imagery to plan and "virtually" fly through missions.

Chapter One
Introduction

By the 2003 Iraqi war, commercial satellite applications had blossomed. For months before the operations began, government agencies were buying commercial satellite imagery every time imaging spacecraft flew over the Middle East. In addition to maps and training for air missions, commercial satellite imagery was used for bomb damage assessment, producing maps for units, looking for weapons of mass destruction, tracking Iraqi units, and following any damage caused by Saddam's henchmen.

But satellite imagery wasn't just the purview of the government and military. It was widely used by news agencies and analysts for following the war. Viewers were given virtual three-dimensional "fly throughs" of Baghdad, Basra, and any other area of interest. News groups followed suspected areas, where weapons of mass destruction might be manufactured or stored, they reported on Saddam's palaces, and pointed out other potential targets with regularly updated satellite imagery.

One such target was Baghdad International Airport. Even as U.S. forces were fighting for the airport, Space Imaging released IKONOS satellite imagery of the deserted airfield that was taken less than 48 hours before U.S. forces arrived on the scene.

While most of the television news groups showed this satellite imagery of the airport, they tended to view it just as "eye candy," something to stimulate the viewer's eyes rather than being a critical part of the story. If, instead, they had looked at the image for news value, they might have noticed a lone Russian transport aircraft parked next to a warehouse. This begs the question, what was a Russian transport doing at a deserted Iraqi airport just hours before being captured by Americans? What were they bringing in or taking out?

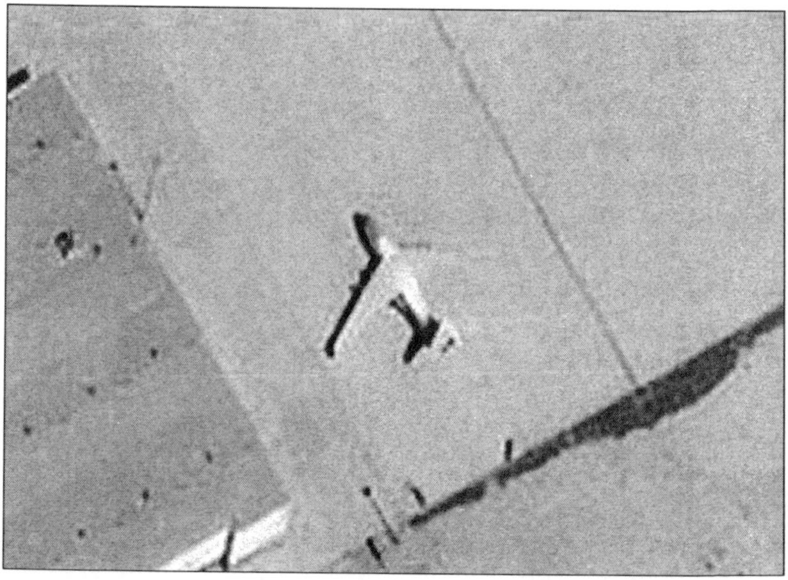

Figure 1.1
What was a Russian transport aircraft doing in Baghdad just two days before the airport was overrun by American forces? (Photo courtesy of Space Imaging)

This glaring mistake by the major news groups proves there is still a place for the average person who wants to make some money in this new, exciting industry. There is a lot of imagery out there, and unfortunately, it isn't being exploited as well as it could be.

Fortunately, prices of satellite imagery have plummeted in the last decade. In the early 1990s, medium resolution French or Russian satellite imagery cost thousands of dollars per scene and took months to get delivery. Today, a better product is available for a few dollars per square kilometer. You can buy it online with your credit card and get a digital image delivered immediately via e-mail.

At the same time, satellite imagery quality has become better. When my book *Satellite Surveillance* was written in 1990, the best available commercial satellite imagery could only see an object as small as 15 feet in diameter. Today, commercial satellites can see objects as small as two feet in diameter! That has opened up new applications that are awaiting someone bold enough to develop them.

While my previous book was written to explain how satellite imagery works, this book is written to show you how to use it for yourself and even make money. Thanks to the proliferation of commercial satellite imagery, the low prices, the availability, and the high quality applications, opportunities for the average person are greater than ever before. Here are some of the areas this book will look at:

Virtual Three-dimensional Satellite Scenes. Ten years ago, this took hundreds of thousands of dollars in software and computers. The only group who could afford to use it regularly was the Defense Department, which used it to help pilots practice bombing runs and special operations insertions.

Today, you need an average desktop computer, inexpensive (or even free) software, and satellite imagery. You can make a computer-generated scene that looks like you are flying over a top-secret installation like Area 51. If you are a police sniper, you can use three dimensional satellite imagery to find the best view of a hostage situation and a clear shot. Or you can go back into history by using spy satellite imagery to recreate a military event from the Six Day War or Vietnam. The only limits are your imagination.

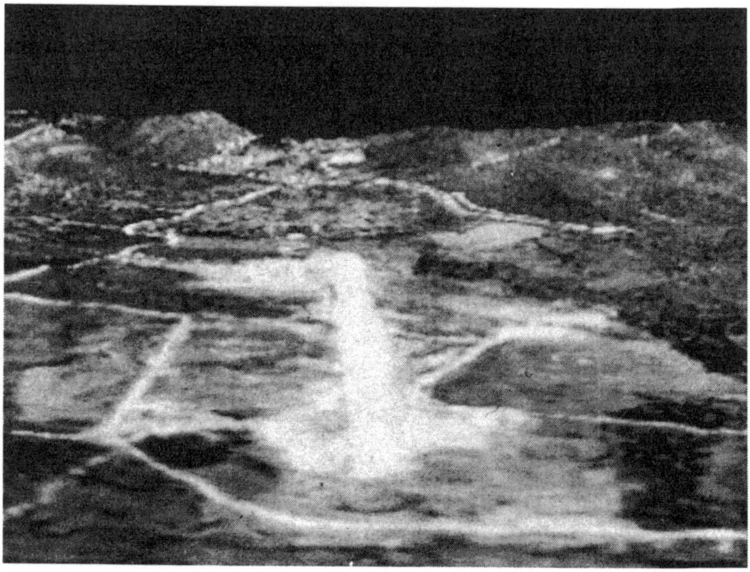

Figure 1.2

Three-dimensional satellite imagery like this was used to train military pilots for special operations insertions. They could virtually fly through an enemy occupied territory beforehand, which lowered the chances of mistakes during the real mission. (Photo courtesy of Autometric)

History. Did you know that the first ten years of spy satellite imagery is available for everyone to look at, providing you know how to find it? This book will tell you how it is stored at the National Archives, how you can copy it with your own camera, and how you can use it to study the Vietnam War, the Cold War, and the 1967 Arab/Israeli War. Game designers can use this imagery to make Vietnam War computer games more realistic. Historians can use it to better understand events.

But this old imagery isn't just for historians. For example, if you are a company that has been charged with recent pollution violations, you would be able to use these photos as solid legal evidence that the pollution existed before you bought the site. We will also look at uncovering declassified U-2 aerial imagery and intelligence imagery from World War Two.

Military Affairs. There are more secret military facilities than photo interpreters to look at them. Just as that Russian transport was missed in Baghdad, a military affairs amateur can find something interesting somewhere in the world.

Commercial Ventures. Everything from mapping towns to real estate, to finding a new location for a new store can benefit from satellite imagery. And, since most people don't know how to exploit it, an average person with a little special knowledge can help make it available to them.

Investigative Journalism. Is Iran completing its nuclear reactors at Bushehr? Where does Syria keep its Scud missiles? Where do the Israelis store their nuclear weapons? What is the state of North Korea's nuclear program? All these stories can benefit from satellite imagery. In fact, in December 1991, ABC News was poised to join the ranks of the great military powers by buying its own high-resolution spy satellite. At the time, ABC executives were in Moscow negotiating with cash-strapped Russian officials for the sale of one of their reconnaissance satellites. Unfortunately, negotiations broke down when the Russians asked for $800 million. As one defense analyst said after learning about the asking price, "The Russians don't really understand capitalism yet."

This book has a complete section that gives journalists or anyone else who wants to do some investigative work, the key to finding critical military facilities. Sample pictures are shown for everything from nuclear reactors, to weapons bunkers, to missiles and ships. Most of these are very recent and many come from recent events in the Middle East. This section may provide just enough of an edge to win a Pulitzer Prize.

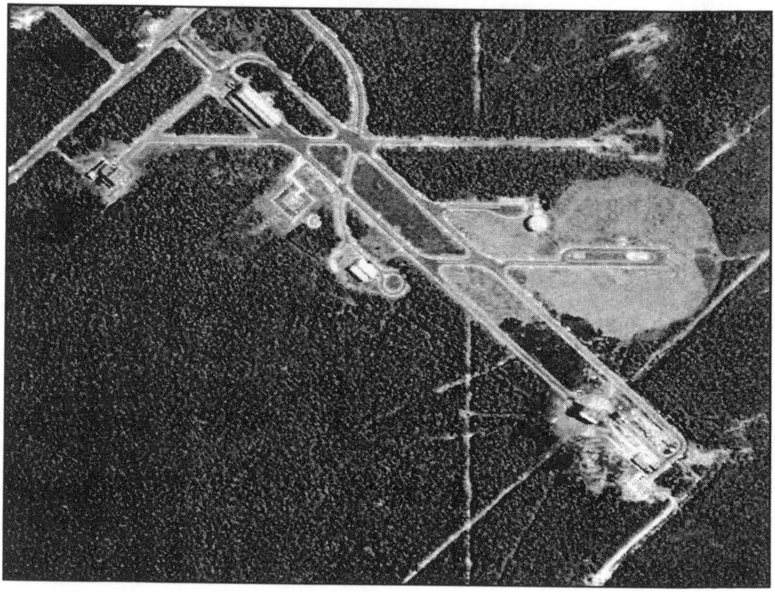

Figure 1.3

The Brazilian space facility at Alcantara. Just days after a rocket blew up on the launch pad in August 2003, this picture was taken by Space Imaging. Note the scarring in the lower right hand corner. As you read this book, you will learn to identify many of the features in this image. (Photo courtesy of Space Imaging)

Prospecting. The major mining companies are using satellite imagery for prospecting. Why not you? If you are one of the many amateur gold prospectors in America, you might want to take a look at satellite imagery. In an afternoon you can cover thousands of square miles looking for likely placer gold deposits, far more than you could ever do in a lifetime of traditional prospecting.

Archeology. Did you know that modern satellite imagery has been used to find everything from Indian villages to sunken ships, and ancient cities? You may not be Indiana Jones, but you can still find some adventure, fame, glory, and even a little money discovering antiques and other collectables in Amcrica's West.

Personal Interests. Is that piece of property you are looking at in a flood plain? Are there potential terrorist targets in your vicinity, and how would an attack on them affect you? Does everyone else in your neighborhood have a swimming pool? Where would be a good place to hunt for quail? A quick look at some satellite imagery can answer your questions.

One final warning. I know that I'm trying to cover a lot of material and leaving out some of the details. My goal is to write a book that everyone, not just satellite experts, can read, understand, and use. If, after reading this book, you can make informed decisions on how to profit from satellite imagery, I was successful. If you want more information on the science of satellite imagery, I suggest you read my book *Satellite Surveillance*. In the meantime, I hope you enjoy reading this book as much as I enjoyed putting it together.

Chapter Two
A Brief History of Satellite Imagery

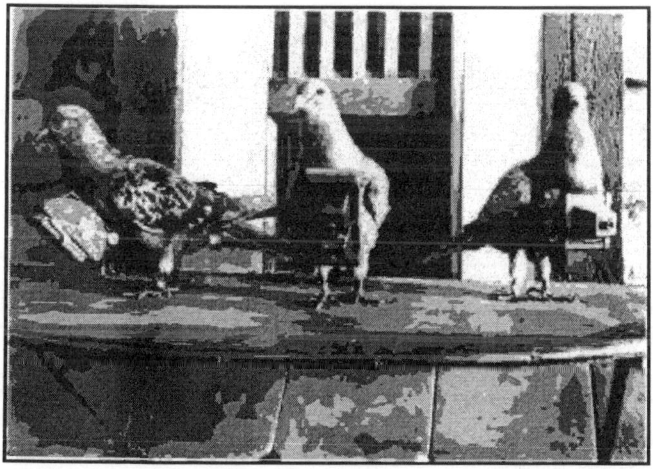

Figure 2.1
Obviously earlier overhead imaging systems were unable to operate effectively in space. (NASA)

Although we could go back beyond the space age to look at the antecedents of satellite imagery (aerial imagery, carrier pigeon photos, etc.) and we could write a whole book on the subject, let's cut to the chase. In this case, we will start with the space race between the U.S. and Soviet Union.

Although the use of rockets as a camera platform was first proposed in 1891, the first photos from space were actually taken in the 1940s, after World War Two. The vehicles for these fledgling space images were confiscated German V-2 rockets, which provided the basis for the Russian and American post war rocket programs. Not only did these missiles provide the foundation of both the American and Russian ICBM programs, they were instrumental in the first studies of outer space.

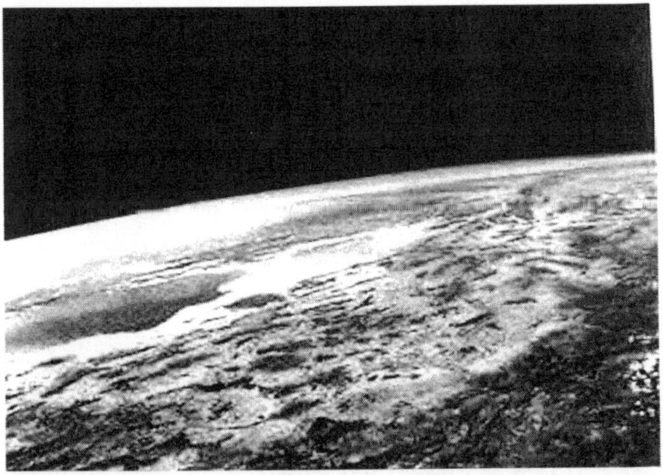

Figure 2.2

America's Viking 12 was launched from White Sands, New Mexico in 1946. This picture shows Arizona and California. (NASA)

The American V-2 program was called Viking and was centered on the 250 V-2s and supporting equipment captured by the U.S. during Operation Hermes. (Operation Hermes was the American operation to get advanced German technology out of Germany at the end of the war.) The Viking project began in 1946 at White Sands Proving Ground, New Mexico. This U.S. Army sponsored project launched at least 62 of the German vehicles (although less than half that number was successful). Of those, some reached into the upper atmosphere and into what is now considered space. On board several of the vehicles were still and motion picture cameras that took photographs as the rocket ascended into space.

As dramatic as the photos were, they had limited use because they were usually oblique shots that only covered areas near the rocket launch site. It wasn't until the late 1950s, when orbiting spacecraft were finally placed in space, that one space vehicle could acquire imagery from many parts of the planet.

Despite the rush by the military and intelligence agencies to get a spy camera in space, it was a civilian application that ranks as the first use of satellite imagery. On April 1, 1960, TIROS-1 (Television Infrared Observation Satellite) was launched into space. Its mission was studying cloud formations and it was the first "real time" satellite, in that it could send its data back to earth on a nearly immediate basis. The problem was that the television cameras of that day had poor spatial resolution and the smallest object the satellite could detect was 100 feet in diameter. This was too poor for traditional photo interpretation, but the Air Force was interested because this was the era of massive nuclear-armed bomber fleets and the Strategic Air Command needed reports on cloud formations and weather reports over the USSR.

Photo interpreters got their satellite imagery with the Discoverer series of satellites. Although billed as a scientific program, Project Corona (as it was known in the intelligence community) was a spy satellite program. The satellites were designed to acquire high-resolution photographs of the Soviet Union and then eject their film canisters into the Pacific Ocean. While the film canister was parachuting down, an Air Force plane would supposedly recover it in mid air.

Launching a satellite and recovering a film canister proved harder than originally thought. The first twelve missions were failures, although Discoverer II did successfully eject its film canister near the Norwegian/Soviet border. Discoverer XIII did succeed, but its payload was a set of sensors that recorded Soviet radar emissions.

Discoverer XIV (Mission 9009) was a complete success and the recovered film canister was rushed to Kodak Labs in Rochester, New York (this would be considered a true Kodak moment). Since film had never been recovered from space before, the experts were concerned that the images might fade soon after developing. Consequently, a CIA photo interpreter was stationed at the end of the developing machine to give a quick interpretation. Fortunately, the images didn't disappear and the interpreters had pictures of the Mys-Schmidta air base and the ICBM base at Plesetsk to study. The resolution was poor, only about 50 feet, but to Andrew Goodpaster, they were, "like the dog that walks on its hind legs, remarkable that it happens at all."

Although the first satellite imagery was poor compared to U-2 imagery, it got better as time went along. When the KH-2 system was launched, resolution was 25 feet. Just a couple of years later, resolution was six feet thanks to the improved optics of the KH-4B system.

While the KH spy satellites were critical to the Cold War, they also laid an important cornerstone to future commercial applications. The KH-3 system was the first to use stereo cameras, which allowed for three-dimensional analysis. KH-5 was primarily a mapping system (ARGON), which was used to develop the accurate maps needed for targeting ICBMs. It set a new standard for map making. And in 1970, the CIA investigated the use of color satellite imagery for mineral exploration.

Despite the successes, there were the glitches and humorous moments like the squashed bug found in the film emulsion, exactly where a secret Soviet facility was supposed to be. There was also the time that the positioning cameras showed a silver sphere in space (no, it wasn't a UFO, but a weather balloon that managed to reach the lower borders of outer space).

Since they were advancing the science of photo interpretation, these photo interpreters for the CIA were soon used for other tasks like analyzing photos of world leaders or secret Soviet equipment. Ironically, one of the biggest tasks that required photo interpretation, analyzing a Russian submarine, which sunk off Hawaii in the late 1960s, was kept from these experts. Although the U.S. Navy had taken photos of the sunken sub with a remote controlled submersible camera, they weren't given to the National Photographic Interpretation Center (NPIC) for interpretation. Instead, Art Lundahl, the legendary head of NPIC created a special taskforce of two interpreters in Maryland who worked on the photos and reported directly to him. No one else at NPIC was aware of the Russian sub or the attempt to recover it from the bottom of the ocean by the Glomar Explorer.

Russian Satellite Imagery Development

Although the United States put spy satellites in space first, it didn't take long for the Soviets to catch up. On April 26, 1962, they launched Cosmos 4 on a Vostok rocket from their launch center at Tyuratam in Central Asia. About a week later, the camera pod came down. The resolution was poor and not good enough to count or analyze American military hardware. However, it could easily see objects like roads, buildings, and rivers, which made it adequate for mapping and targeting strategic American targets.

What followed was a highly successful spy satellite program that evolved differently than the U.S. spy satellite program. Since U.S. society was open, the Soviets didn't have the critical need for satellite imagery, so there was less priority on the project. Their satellites were larger and less miniaturized since the Russian lead was in large boosters. This also gave them the opportunity to make use of world-class film and lenses from their East German allies and Zeiss of Jena.

The second generation of Soviet spy satellites were much improved and launched from Plesetsk into a polar orbit. Ironically, the U.S. was unaware of their purpose until a British schoolteacher, who tracked satellite orbits for a hobby noticed their different orbit and determined their purpose.

These new satellites fell into three categories, high resolution, low resolution, and maneuverable satellites. While the low and high resolution satellites stayed up for eight days, the maneuverable satellites remained in space for 12 days. This generation of satellites continued to stay in use until the Fall of 1983, when the third generation satellites were launched. The new class of satellites were all maneuverable and remained in orbit for 12 days. However, they were separated by the optics: high, medium, and low resolution.

Although Soviet photographic systems were world class, they wanted electronic imaging that gave them real time imagery. However, they had fallen far behind the U.S. in electronics by the late 1970s. Since photographic imagery still offered the best resolution, the fourth generation spy satellite still produced high-resolution photographic products, although a low resolution TV camera was onboard. In order to get up-to-date information, they relied on frequent, timely launches of their spy satellites.

In order to keep a lead in resolution, the Russians used the highest resolution black and white film. They obtained color images by using a bank of cameras and special filters that looked at specific bands of light. The negatives could later be matched to provide "color" images.

Russia finally solved the electronics problem on March 29, 1984, with the launch of Cosmos 1546. It was the Soviet equivalent of the KH-11 in that it gave "real time" imagery. However, it reflected several compromises due to the fact that the Russians had fallen behind in both the space and electronics race. First, since it wasn't launched by a major rocket booster like the American Space Shuttle, the satellite was smaller. And, since Soviet solid-state electronics weren't as good as American electronics, the satellite's capabilities were limited. It only had a one-dimensional array of light detectors, which collected imagery as it swept across the terrain, much like a broom sweeps over the floor. It also had fewer shades of grey in order to limit the amount of data that had to be relayed back to earth.

Cosmos 1546 was maneuverable and could change its orbit from 200 to 400 kilometers in altitude. It had a focal length of 1.375 meters and a resolution of about 2 meters. In order to maximize resolution, it only transmitted black and white images.

The Soviet satellite had many weaknesses. The width of the scene was only 16 kilometers, which kept it from looking at large areas. The poor Soviet space communications capabilities meant they were unable to transmit as much data. Like its American counterpart, the Soviet satellite had a thermal infrared sensor, which could see in the dark, but it was considerably worse than the American system, which meant the Soviets were usually blind when their satellites traveled over the dark side of the Earth.

The Soviets continued to rely on photographic systems. The highest resolution imagery was still provided by traditional cameras, while low-resolution cameras, calibrated especially for mapping, photographed large areas.

Although the Russians remain behind the West in satellite imaging technology, the end of the Cold War has allowed them access to more Western technologies. As we will see in later chapters, the Russians are still in the satellite imagery business and a good source for some imagery, especially high quality photographic products.

The Evolution of Commercial Satellite Imagery

Commercial satellite imagery has always been the poor stepsister of the military reconnaissance systems. Since space was an expensive endeavor, few companies could afford the high cost of launching a commercial imaging satellite, especially since their capabilities were severely limited by the intelligence community. That's why most of the 100 or so commercial satellites are involved in communications.

The decision to restrict commercial satellite imagery was made in the early 1960s, when President Kennedy created an interagency committee to study the political aspects of America's spy satellites. One of the suggestions of the committee was to limit civilian satellite spatial resolution to no less than 98 feet. The recommendation was accepted and for several decades the U.S. LANDSAT satellite program was limited to that resolution. It wasn't until the 1990s, when the French and Russians began to dominate the space imaging industry, that American companies were allowed to compete.

Ironically, the limitations on resolution forced the commercial satellites to focus on what is called multispectral imagery. Rather than just look at visible light (red, green, and blue), the sensors began to look at other parts of the electromagnetic spectrum like near infrared and thermal infrared (for more information on multispectral imagery, please refer to my book *Satellite Surveillance*). These different bands of electromagnetic energy were discovered to have special applications in earth sciences, so the U.S. initiated the Earth Resources Technology Satellites Program (ERTS) in 1967. The first satellite, ERTS-1 was launched in 1972, onboard a Nimbus weather satellite. The success of the ERTS program led to the launching of several more satellites, now called LANDSAT.

The success of multispectral imagery had an effect on the American intelligence community. Several voices in the CIA argued for more spy satellite missions with color and infrared film because they could detect objects that were invisible to traditional panchromatic film. Unfortunately, these voices were usually in the minority and most missions continued to use higher resolution black and white film, which usually had about twice the ground resolution.

In the early days, civilian satellite images were used nearly exclusively by government agencies and the scientific community. Private industry was slow to use satellite imagery because few business leaders could understand the imagery and the often complex science behind it. For instance, when geologists began to show mining executives that multispectral imagery could aid in mineral exploration, they were skeptical. Although executives understand cash flow and return on investment, they were lost when a geologist tried to explain how looking differently at red, green, blue, and infrared light can find a mineral deposit. It didn't help that satellite imagery was expensive at the time, too.

By the 1980s, the American monopoly on satellite imagery was breaking up. In 1986, the French launched the SPOT satellite (Satellite Pour l'Observation de la Terra). It could see visible light and near infrared. Best of all, it had a spatial resolution of 10 meters, which opened up more civilian applications.

Ironically, the end of the Cold War, which inspired the race for space imagery in the first place, brought in a new player: the Russians. Post-Soviet Russia, which was in desperate need of hard currency, decided in the early 1990s to sell the product of their spy satellites. At first, they offered 10 meter resolution and 5 meter resolution imagery from two of their reconnaissance satellite systems. Later, they changed the standard of high resolution satellite imagery by offering 2 meter imagery from their KVR-1000 satellite system.

By now the floodgates on space imaging had opened up wide. Soon, other countries like Japan, Brazil, India, the European Community, and Canada started putting sensors in orbit.

Many times they were just part of the payload on a space shuttle mission, other times they were a separate satellite. But, however it was done, there was a growing archive of satellite imagery of the Earth.

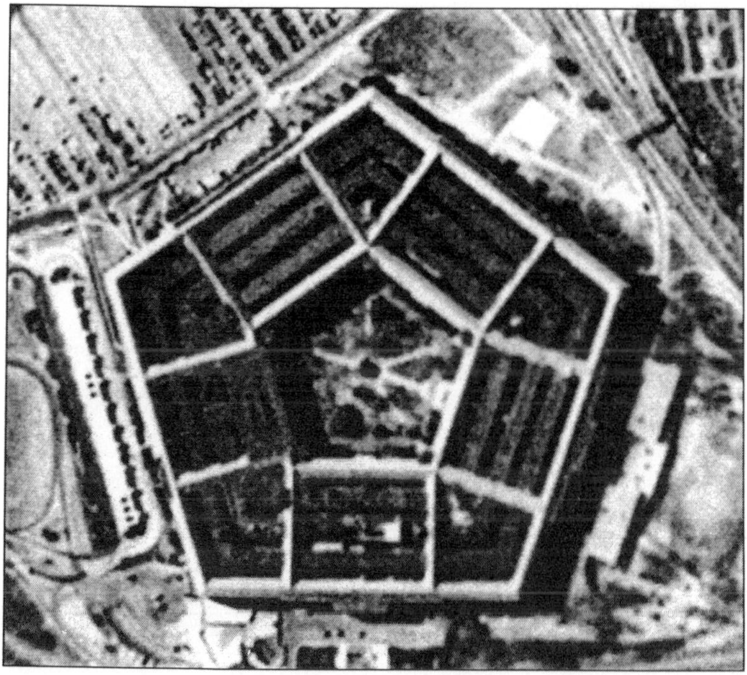

Figure 2.3

The first high resolution imagery reaches the civilian market. This two-meter resolution image of the Pentagon was a sample circulated by the Russians to interest potential clients. (Sovinformsputnik)

It was obvious that the U.S. government couldn't control the resolution of satellites in space any longer since so many countries, especially the Russians and French were in the business. As a result, the U.S. government deregulated the industry, while maintaining several restrictions concerning national security and satellite coverage of Israel. This allowed the U.S. space industry to do what it does best and in 1993, the U.S. Department of Commerce granted the first license to build and operate a high spatial resolution digital imagery satellite for commercial purposes.

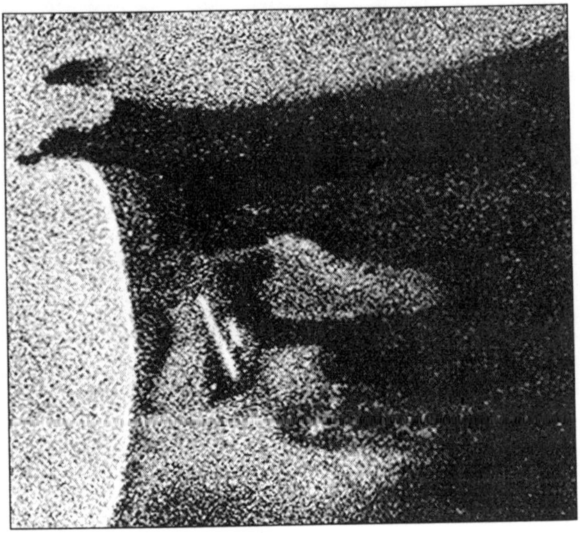

Figure 2.4

The pioneer of commercial satellite imagery. This image of a Soviet airfield at Mys Shmidta is the first image taken from an orbiting satellite. (Central Intelligence Agency)

The first satellite to successfully take advantage of this new policy was IKONOS, which was the first commercial satellite with one-meter resolution. The 1,600 lb. satellite orbits the earth every 98 minutes at 423 miles. The orbit is sun-synchronous, which means that it passes over about 10:30 am local time. This is early enough in the day that heat doesn't degrade the atmosphere too much. It also offers some shadows so analysts can measure the heights of objects on the ground.

However, the race for the best space-based sensor isn't over. The best spatial resolution satellite, as this book is being written is Quickbird 2, which was launched in October 2001 and boasts a resolution of 2 feet. And, lest multispectral imagery is forgotten, scientists are looking at hyperspectral imagery, which records hundreds of distinct parts of the electromagnetic spectrum. This will be a boon for earth resources scientists, because they will be better able to determine the exact chemistry of an area, just by studying how light interacts with the various chemicals in the ground.

The growing satellite business also had a beneficial effect on the U.S. government's obsession with secrecy. When it became obvious that the world wasn't going to fall apart with the availability of high-resolution satellite imagery, the Clinton Administration ordered the declassification of imagery from the nation's first spy satellite program (CORONA). Today, anyone can go to the National Archives outside Washington, DC, or the EROS Data Center in Sioux Falls, South Dakota, and acquire this top secret satellite imagery that was so critical during the Cold War.

Summary

Commercial satellite imagery has advanced dramatically in the last fifteen years. Unlike the first 25 years of space imaging where the military and intelligence communities benefited, most of the recent improvement has come in commercially available imagery, which has finally been released from the artificial restrictions of intelligence guidelines. Today, there is a wide assortment of satellite imagery for a wide range of prices. And, as we will see later in this book, there are many applications that have hardly been touched.

Civilian Satellite Imagery Programs

This is a list of civilian satellite imagery programs that have actually successfully launched a satellite. Some like LANDSAT and RESURS have launched many vehicles over the years. At the same time, the imaging systems have improved, which has allowed better spatial resolution and coverage of more spectral bands. No wonder keeping track of satellite imagery can be confusing.

Program	Country
LANDSAT	U.S.
SEASAT	U.S.
HCMM	U.S.
RESURS	Russia
IRS	India
ERS	Europe
JERS	Japan
RADARSAT	Canada
ADEOS	Japan
TERRA	U.S.
Nimbus	U.S.
SPOT	France
Orbview	U.S.
SPIN-2	Russia
IKONOS	U.S.
Quickbird	U.S.
EROS	Israel

Chapter Three
How Satellite Imagery Works

In the last chapter you read the phrase, "spatial resolution." You may have even wondered what it meant. Well, thanks for hanging in there because we will now talk about resolution.

The normal definition of resolution is the smallest object something (either a camera, photographic film, or spy satellite) can detect. In the case of the last chapter, when we talked about a satellite having a resolution of one meter, the smallest object the satellite could detect was one meter in diameter. Note that I said "detect" not identify. The object could be a pile of dirt, or an engine block, but you couldn't tell from the satellite image. Of course, the surroundings can give you a good hint. For instance, if you were viewing a construction site, it would be more likely to be a pile of dirt than an engine block. Of course, it could just as easily be a stack of bricks, too.

Now that I've given you the easy definition, let's make it complicated. For the purpose of this book, we will talk about two types of resolution: spatial resolution and spectral resolution. Spatial resolution, as we just talked about is the smallest object that you can detect. Spectral resolution involves ability to detect smaller segments of the electromagnetic spectrum.

Here's an example of the differences between spatial and spectral resolution. Kodak produces a black and white film called Panatomic X. It is used by professionals who want the highest spatial resolution because they want to enlarge their photographs. It can differentiate up to 300 lines per millimeter (meaning that it can record something as small as .0033 mm in diameter on the negative). However, it has poor spectral resolution because it records visible light as either black or white; there is no differentiation between greens and reds.

A color film, on the other hand, may have a worse spatial resolution of .01 (one hundred lines per millimeter), but will record red, green, and blue, and also many shades in between. What it loses in spatial resolution, it more than makes up for in spectral resolution.

The same is true with satellite imagery. Just as some want the maximum resolution and are willing to trade color for it, there are some who will need the highest spatial resolution satellite imagery available, even if it is just black and white (called panchromatic by professionals). Others need more spectral resolution because they need to study objects like plants where the spectral information is critical.

With these basic definitions in mind, let's continue.

Understanding Spatial Resolution

Despite all the previous talk about the importance of spectral information, the benchmark for imaging satellites is still spatial resolution. When a new satellite is launched everyone says, "I hear that SuperBird I was launched yesterday. It has a resolution of 45 centimeters." I have yet to hear anyone say, "I hear that SuperBird I was launched yesterday. It has a spectral resolution of one nanometer in the shortwave infrared band." Well, okay, there are probably a few nerdy scientists who get excited about things like that, but they are few in number. In reality, satellite spatial resolution is like a processor speed or accelerating from 0 to 60 mph in x number of seconds for cars; the standard everyone tries to better.

Unfortunately, there are a lot of factors in spatial resolution that you have to understand. The final resolution figure that the satellite manufacturers brag about is a function of the sensor size, optics, and the satellite's orbit.

Lest you think that these new high resolution satellites are something special, let's compare them to readily available camera equipment. The formula for determining resolution is: Resolution = (Distance/Focal Length) x Resolution diameter. Resolution diameter is the size of the light-sensing element in the system, whether it's a charged couple device (CCD) or the grain size of the silver compound in the film. In the case of high resolution professional films, the resolution is over 300 lines per mm.

Using the formula above, let's see what the resolution of a satellite could be if we used readily available camera equipment. If we put a 35mm camera on a satellite that had an orbit of 150 kilometers and used a commercially available 1,000 mm lens and Kodak Technical Pan film, the equation would look like this: (150 km/1,000mm) X .000003mm = .45 meters. That means commercially available equipment costing about $1,000 could see something the size of a newspaper on the ground.

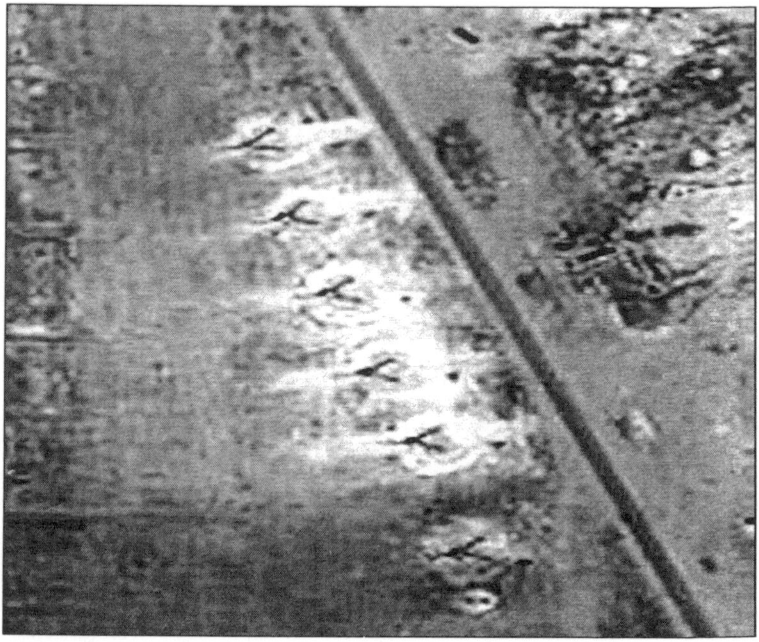

Figure 3.1

How good is one-meter resolution satellite imagery? Here is a battery of Syrian Howitzers near Homs, Syria. (Space Imaging)

I know what you are thinking. If commercial cameras are so good, why do we have expensive spy satellites? We can just send up a camera with the astronauts and develop the film at the local Wal-Mart. Good question. The fact is that many manned space missions are primarily Department of Defense missions and the astronauts do use commercially available cameras to take snapshots out of the window. However, keeping an astronaut up in space with a 35mm camera is much more expensive than a fully automated satellite imaging system.

Of course, there is more to acquiring high spatial resolution imagery from a satellite than picking up some camera equipment from the local mall. Obviously, distance from the object means a lot. The 150 km orbit used in this example is quite low and the satellite would reenter the atmosphere in a short time. However, that is what the early spy satellites did. Today's imaging satellites cost hundreds of million of dollars and in order to recoup the investment, long orbital lives, and subsequently higher orbits are required.

Another factor in resolution is the size of the light-detecting device. Obviously, high-grade photographic film has a smaller size grain than the smallest CCD. Of course, silver film can only be used once, while CCDs can be reused again and again. In addition, while photographic film only uses about 2 percent of the light that strikes it, a CCD can convert over 80% of the light into a signal. That means it can be used in low light, which explains why a modern video camera or digital camera doesn't need additional light in dim light situations where a conventional camera would need a flash attachment.

There are also other factors, not mentioned by the satellite manufacturers that have an impact on the final resolution of a scene. For instance atmospheric conditions like smog, humidity, and cloud cover will obviously have an effect on what you can see. And, there is contrast. While a small mirror lying on the ground on a sunny day will be easy to find from space, because it stands out from the background, a polar bear in a snow field will be hard to detect with a high spatial resolution satellite imaging system. That's why spectral resolution is so important.

Spectral Resolution

While a high spatial resolution camera may have problems finding that polar bear on a snow field, the high spectral resolution system will have an easier time. For instance, while snow is cold, the bear, being a warm-blooded mammal will actually radiate heat and be detectable to a thermal infrared sensor. Snow is also made up of ice crystals, which also reflect more light than the bear's fur, so although both are white, the snowfield will reflect more sunlight than the bear's fur.

Humans also use spectral resolution in everyday events. For instance, you enter the kitchen and see the electric coils on your stove glowing a dull orange. You immediately know that the element is hot because you know metal that glows orange is HOT! You then go over and turn the element off. Several minutes later you walk back in and see that the element is gray once again. However, instead of precipitously touching it, you wave you hand a couple of inches from the element to see if you can feel any radiated heat.

You have just used two spectral sensors and two parts of the electromagnetic spectrum. The first sensor was the eye, which told you that the heating element was hot enough to glow orange. That's using the visible part of the electromagnetic spectrum. When that sensor system didn't give you enough information a few minutes later, you employed your second sensor, the hand. That is using the thermal infrared portion of the electromagnetic spectrum.

For all the talk about spectral resolution, multispectral imagery, and hyperspectral imagery, scientists use the information and sensors in the same manner that you did in the kitchen. You determine the characteristics of what you are trying to detect and use the sensors that you have available. In most cases, you are looking at something on the ground and determining how it reflects sunlight.

If you are trying to determine the health of a farm crop, you want to use the near infrared band because vegetation reflects solar infrared energy, especially when it is healthy. If the amount of infrared light being reflected were lower in one field than in others of the same crop, it would make sense to check out the field to see if there are any problems. Ironically, the plant's reflectance in the shortwave infrared region increases as it becomes ill.

If you have a satellite imaging system with a high spectral resolution, you can learn even more. Each type of plant has its own reflective pattern, and the plant's reflective pattern changes as the plants mature. With a higher resolution spectral image, you can better determine the type of crops, how well they are maturing, and when they will be ready for harvest.

High spectral resolution is also critical in mineral exploration. Although many minerals and rocks look alike to us, they reflect sunlight in radically different ways. With computer enhancements, geologists can identify different families of minerals. In some cases, the mineral may affect plants, so by looking at the infrared reflection of the plants on the ground, geologists can gain insight into the geology.

"So if high spatial resolution and high spectral resolution are both so good, why isn't there a satellite that offers both?" you ask. Good idea. However, physics limits what we can do.

There is a trade-off. What we can sense depends on how many photons (light particles) we can collect and detect. If you want high spatial resolution, you want a small sensor (remember the equation). The problem is that small sensors cover a smaller area, so they will collect a fewer number of photons.

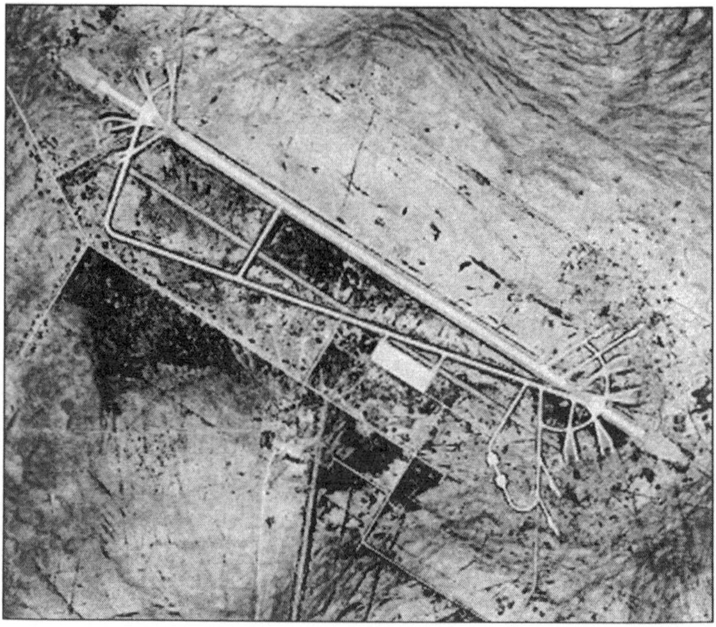

Figure 3.2

You don't need high resolution imagery to see that this is a military airfield. The revetments at each end of the runway are to protect fighters from air attacks. This airfield is in Western Iraq and was captured by U.S. Special Forces to prevent Scud attacks on Israel. (Space Imaging)

Suppose you want to build an imaging system that can gather high spatial and spectral imagery; one-meter spatial resolution and 20 bands of spectral information that covers everything from blue to near infrared. Sounds great! The problem is that in order to collect enough light to register in all twenty bands you need to collect more photons (a photon in the green wavelength, one in the red wavelength, etc.). Yet the size of the sensor is limited in order to keep the high resolution. It's like catching rainwater. You can use one big pail and catch a lot of rain or you can use smaller pails and expect to get a smaller amount in each pail. You can't use smaller pails and expect to get as much per pail as you did with a larger pail. The principle is the same as that of the human eye. In dark conditions, where there are fewer photons, the eye only detects black and white with its rods. Color vision is reserved for brighter conditions and the cones in the eye.

Most satellites have two types of sensors (just as the eye has cones and rods light sensors). One is a panchromatic sensor that collects all of the light and gives a high spatial resolution panchromatic (black and white) image. The other is a multi-spectral sensor that looks at several distinct parts of the electromagnetic band, but has poorer spatial resolution. When they are retrieved from the satellite, they are often blended by computer to give a color image that has the high resolution of the panchromatic scene. They are quite appealing to the eye, but the process of blending them does cause a loss of information. That's why professionals who rely on high spectral resolution want the separate bands of information.

Picking the Right Imagery for Your Purpose

If you feel confused, don't worry. There are a lot of satellite professionals who are confused by the rapid proliferation of spatial and spectral satellite imagery. However, you needn't be totally familiar with each type of sensor on every satellite. You just need to know what you need for your application. The fact is, several systems may meet your needs, and your final decision should be based on cost, availability of the information you need, timeliness and the spectral resolution required for your purpose.

Although every satellite imaging system is slightly different, the sensors still fall in several large categories. Here they are:

Panchromatic. These are the high resolution satellite images that need to gather every photon of light to achieve the best resolution possible. Although each satellite's system will differ slightly, they generally collect all the light from the blue, green, red, and near infrared parts of the electromagnetic spectrum. (See Figure 2.3 on page 21 for an example.)

These panchromatic images are very popular with anyone looking at manmade objects or who are interested in civil engineering projects. They are less important in earth resources studies because they don't have the spectral information needed. However, they are often blended with multispectral imagery to give a sharp, color image.

Generally, the only reason for using panchromatic satellite imagery is to get the best resolution possible. Consequently, the poorer the resolution, generally the lower the price. If you don't need the best possible resolution or want to cover a larger area (generally the lower the spatial resolution, the larger the area covered), the best bet is to go with a lower resolution, cheap satellite image. For instance, if you want to make a road map of an area, it would make more sense to get a two or three-meter resolution image that will show roads, rather than a more expensive image that has two-foot resolution.

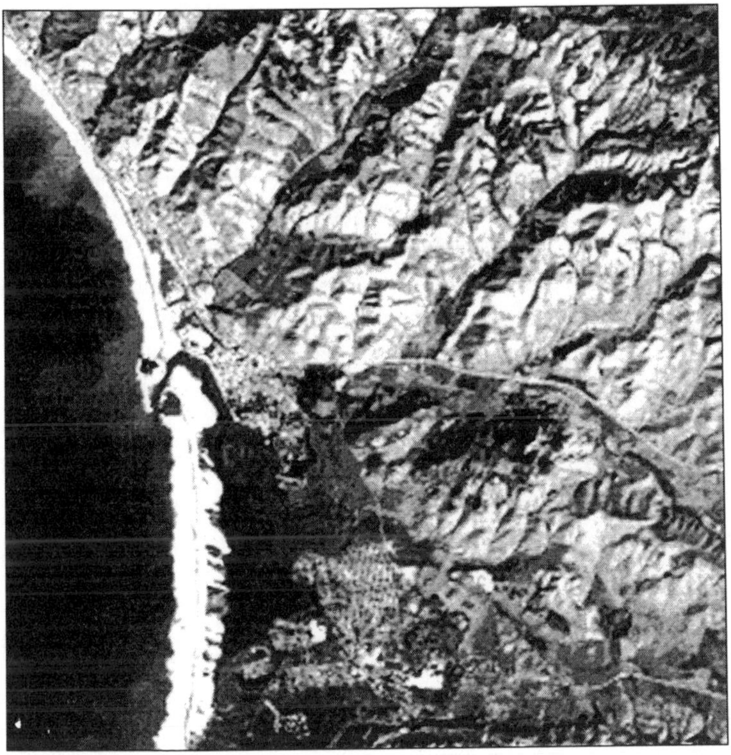

Figure 3.3

This is a panchromatic image showing green light from a coastal town. Note that water less than 90 feet deep is seen as a grey. (NASA)

Visible Light. This is the part of the electromagnetic spectrum that has a .4 to .7 micron wavelength. It's the only group of light that can see underwater and tends to scatter more than other bands like infrared. It is obviously broken down into three basic colors, red, green, and blue.

Blue Band. .4 to .5 micron wavelength. Blue light can penetrate up to 150 feet of water, so it is an excellent band to study deep water. Of the three visible bands of light, it is the most likely to scatter in the atmosphere, which limits detail.

Green Band. .5 to .6 micron wavelength. Green light can penetrate up to 90 feet in clear water, so it can be used in conjunction with blue band light for water studies. It is also good for any project that looks at vegetation.

Red Band. .6 to 7 micron wavelength. This band of visible light scatters less than either green or blue, which provides more detail. It also penetrates up to 30 feet in clear water, so it can be used in conjunction with blue and green bands for underwater studies. Since it is the clearest visible band of light, it usually provides the best visible spectral information. (See Figure 6.3 on page 118 as an example.)

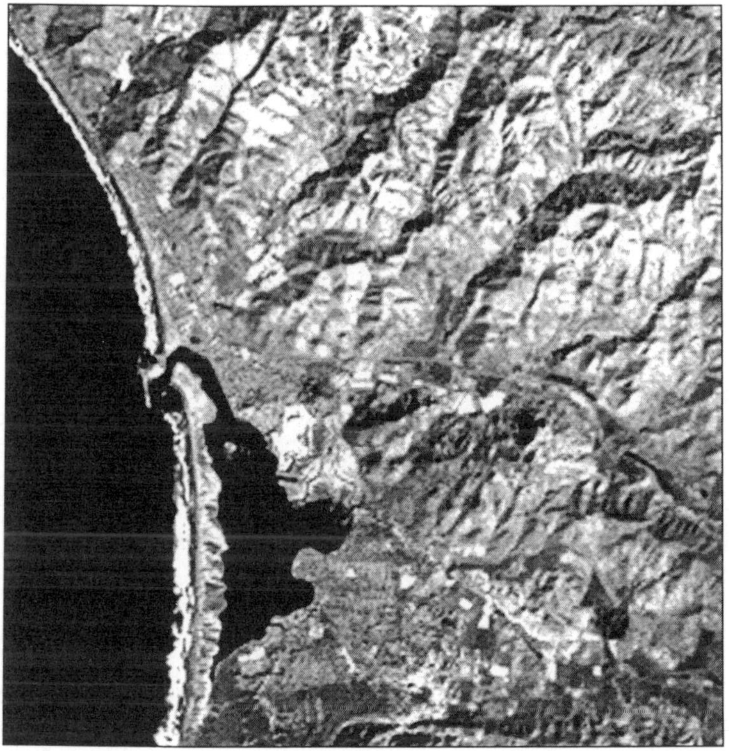

Figure 3.4

This near infrared image of the coastal town doesn't show any shallow water since water absorbs near infrared energy. The land is lighter, however, because healthy vegetation is reflecting the near infrared light. (NASA)

Near Infrared. .7 to .9 microns wavelength. This is the part of the electromagnetic spectrum that is best for vegetation studies because healthy plants reflect large amounts of near infrared energy back into space. This band is obviously important to finding camouflage.

Middle Infrared (also called shortwave infrared). 1.5 to 2.35 micron wavelength. This band can detect moisture content in ground, so it is excellent for real estate development, civil engineering, and transportation projects. Military planners use this band to see what problems they may have in moving heavy tracked vehicles like tanks across open terrain. This band can also detect hot objects like fires.

Thermal Infrared. 10.4 to 12.5 micron wavelength. This shows emitted heat energy instead of reflected energy. It can be used to detect water currents, geological structures, and manmade, heat-generating sources. Since the resolution is poor, its applications are limited.

Unless you are going to specialize in the select field of hyperspectral imagery, these general classifications should hold you in good stead. As a result of this general classification, there are only a few types of images that provide the most spatial and spectral information in an image. Some take one band each from visible, near infrared and middle infrared since that gives the broadest amount of information in one color image. However, the most popular imagers are:

True Color Image. This is the most natural image for anyone not used to viewing satellite imagery. It looks like what one would expect to see from a plane and is good for showing to people unfamiliar with satellite imagery. Since the evolution in high spatial resolution imagery, many satellite imagery providers blend their high spatial resolution imagery with a lower resolution color image to give the best of spatial and spectral resolution. (See Figure 4.1 on page 52 as an example.)

Infrared Image. This is a popular color image in vegetation studies. In it, one of the primary colors (blue, green, or red) is eliminated and infrared is substituted. Traditionally infrared has been displayed as red since red is closer to infrared and is often associated with heat. However, since plants reflect infrared, many specialists use green for infrared because it highlights stands of vegetation. The choice is up to you. (See Figure 5.1 on page 66 as an example.)

If you aren't studying manmade objects or vegetation, the choice gets a bit harder. Here is a table of the best type of spectral bands for studying various features.

FEATURE	**SPECTRAL REGION**	**COMMENTS**
Clear water	2.08 to 2.35 microns	This band of energy is absorbed by water and gives a clear demarcation between land and water.
Croplands and pastures	1.55 to 1.75 microns	Medium grey in this Band.
	2.08 to 2.35 microns	Light tone in this band.
Faults and fractures	1.55 to 1.75 microns	Shows linear topography.
Grasslands	1.55 to 1.75 microns	Light tone in both bands.
	2.08 to 2.35 microns	
Deciduous forests 1.	55 to 1.75 microns	Very dark in this band.
	2.08 to 2.35 microns	Light in this band.
Coniferous forest	1.55 to 1.75 microns	Very dark in this band.
	2.08 to 2.35 microns	Mottled medium grey in this band.
Defoliated forest	1.55 to 1.75 microns	Lighter in this band than next one.
	2.08 to 2.35 microns	Darker than pervious band.
Mixed forest	.76 to .9 microns	Blotchy grey tones.
	2.08 to 2.35 microns	
Moist Ground	2.08 to 2.35 microns	Irregular darker grey tones.

FEATURE	SPECTRAL REGION	COMMENTS
Nonforested coastal wetlands	2.08 to 2.35 microns	Clear demarcation between land and water
Silty water	.76 to .9 microns	Will delineate waters edge
	2.08 to 2.35 microns	Water will be dark in this band
Sand and beaches	.76 to .9 microns	Bright in all bands. Bands will
	1.55 to 1.75 microns	delineate water. Use also with quarries.
Underwater studies	.4 to .5 microns	Can see bottom in up to 150 ft of water
	.5 to .6 microns	Can see bottom in up to 90 ft of water
	.6 to .7 microns	Can see bottom in up to 30 feet of water
	.76 to .9 microns	Clearly marks water line

If you need to study these types of features, you will be forced to rely on LANDSAT, IRS, or ASTER data, because they are the only systems that give more information in the infrared regions. Most other systems focus on visible and near infrared light because that is the region of the electromagnetic spectrum that CCDs work best in. For more information on multispectral imagery and how it is manipulated, please read my book, *Satellite Surveillance*.

Computers and Software

In the early 1990s, the computer and software usually cost more than the satellite image. Imagery came on mainframe magnetic tapes, top of the line Silicon Graphics computers were required to process the imagery, and software ran into several thousand dollars per copy. No wonder prints of satellite imagery were popular.

Today, you can still get a print of a satellite image, but most personal computers are more than capable of handling the large files and basic enhancement techniques. Consequently, if you have any serious interest in satellite imagery, use your home computer or shell out the few hundred dollars for a basic desktop computer. Take it from someone who has used both prints and computers — prints are a pain (and hard on the eyes).

There are several reasons for using a computer. First, the data from satellites is digital, so you are receiving information that is the closest to what the satellite actually sent. Second, many satellite image providers have web sites that can automatically send your digital image via e-mail once you pay online. No more waiting days or months for the image. Third, a computer screen is easier on the eyes when looking at imagery, especially since you can zoom in on an area of interest. Fourth, there are a lot of satellite imagery web sites that you can visit online (some like Space Imaging even offer free samples that you can play with on your computer to get practice). Finally, most basic digital photo software found on computers today can handle many satellite imagery enhancements (if you get into multispectral and hyperspectral imagery, a better suite of software will probably be required).

If all you have is a basic computer, you can handle most of the satellite imagery you will probably want to use. High resolution satellite systems like Ikonos, Quickbird, and SPOT all can send their images in popular electronic formats like tiff and jpeg. In many cases, these images are already manipulated so little if any electronic massaging is needed. Just load the image into your favorite image software, and you can trim it, manipulate contrast, change tones, or whatever you want to do to make it a better product. In fact, current software is so good that 90 percent of my satellite imagery work is done with the free software I got with my computer. And, in most of those cases, it is just as good or better than the software that satellite specialists paid thousands of dollars for several years ago.

There are times, however, when cheap software can't do it. If you wish to merge several bands of imagery to create one color composite, or the satellite image format is designed for remote sensing, or you want to do more sophisticated analysis, you may have to shell out between $100 to $2,000. But, before doing that, check out some more sophisticated photo enhancement software found at the local computer or office supply store. Common photo software like Adobe can do wonders.

If these can't help, then the only solution is to buy some real satellite imagery software. Although there are many on the market, one of the best is ENVI (Environment for Visualizing Images). This state of the art image processing system is designed for satellite and airborne imagery and is used by scientists and the intelligence community. It can handle the latest multispectral and hyperspectral imagery and make it do handstands.

Computer software is also available that can take satellite imagery and make three-dimensional perspectives with it. We will cover that subject more in a later chapter.

So now you know the difference between spatial and spectral resolution. You know what parts of the electromagnetic spectrum are best for looking at different features. If you have a computer and have managed to download some satellite imagery to play with (if you haven't, try www.spaceimaging.com

and go to the gallery), I guess it's time to see what you can do with satellite imagery.

Chapter Four
What Satellite Imagery Can Do

So far, we have talked a lot about the theory of satellite imagery, but little on how you can use it in your own life, either to further an interest or to make money. With that in mind, let's look at how some others have successfully used it so you can have a better idea of its many applications.

Searchers of The Lost City (Archeology). It's the type of adventure that Indiana Jones would love; a fabulously rich city lost for nearly 2,000 years after mysteriously disappearing. It's a story interwoven with deadly vipers, mercenaries, and ancient curses. However, in this case, Indiana Jones would need to trade his whip and hat for a desktop computer and a digital satellite image; the tools several archeologists used to find the ancient trading city of Ubar on the Arabian Peninsula.

According to legend, Ubar was a rich, beautiful city built as an imitation of paradise. The Koran called it a place of, "lofty pillars," the like of which were not produced in all the world. It was also the source of many Arabian traditions like Aladdin and Ali Babba's cave.

Ubar's wealth wasn't due to anything magical, however. The town was the center of the frankincense trade. The bushes which produced the fragrant gum were grown about 50 miles to the south and caravans from all around the known world came to Ubar to purchase the expensive, crystallized gum. It's even possible that the frankincense given to the Baby Jesus by the Wise Men came from Ubar.

But all of its wealth didn't save it from a dramatic death around 300 A.D., when the Earth swallowed up the city, the sands covered it, and it disappeared. The Koran said it was a city of sinners that was destroyed by God, like Sodom and Gomorrah.

Every great discovery needs someone with the sole of an adventurer and Ubar had Nicholas Clapp, an Emmy winning filmmaker, who learned about the city when reading a book about a British explorer. He became interested in using satellite imagery, when he heard how Space Shuttle radar had discovered ancient riverbeds under the sands of Egypt. He wondered if something like that could find Ubar.

Clapp managed to convince NASA's Jet Propulsion Laboratory to help him and in 1984, the Shuttle's SIR-B flew over the Saudi desert in an unsuccessful attempt to find the city. They then tried LANDSAT imagery, but without success. However, rather than quitting like many would do, Clapp began looking at parts of the Arabian Peninsula that weren't considered likely locations for Ubar.

One of these scenes showed, in addition to the bright reflections of modern gravel roads, a faint network of ancient caravan trails coming together at Shisar. These trails were the paths taken by thousands of animals carrying their precious cargo to the remote parts of the known world like Rome, Jerusalem, and Greece. The reason they hadn't noticed it before was that the site wasn't in the Empty Quarter as originally believed, but just to the north of it. Nor, was it ever covered by sand dunes as proclaimed by legend. When scientists went to Oman to brave desert mercenaries, scorpions, and vipers, they discovered a city that mirrored the descriptions of Ubar in the legends. It appeared that as the town grew, it used more water and finally the limestone reservoir collapsed under the weight of the city, plunging the city into the earth.

In the movie *Indiana Jones and The Last Crusade*, a scholarly Indiana Jones tells his archeology students that X never marks the spot and there aren't any treasure maps. However, with the aid of satellite imagery, archeologists have the closest thing to a treasure map. Of course, finding the fame and fortune and adventure is still up to them.

Controlling Forest Fires (Forestry Management). A forest fire firestorm is just as dangerous as a nuclear explosion. Temperatures soar over a thousand degrees and wind velocities exceed 200 miles per hour. Any unfortunate human or animal caught by it dies from asphyxia as the oxygen is sucked from its lungs.

Forest fires aren't just a threat to fire fighters and wildlife. Within minutes a firestorm can wipe out valuable lumber and vegetation that took centuries to grow. They can also destroy local economies dependent on tourist dollars.

For over the last decade, forestry officials have closely tracked fires with the help of satellite imagery. Thanks to the thermal and medium infrared bands on LANDSAT, they can see the fires, even when the scene is covered in smoke, while other types of satellite imagery help them access damage and future threats.

Satellite imagery can do more than just monitor a forest fire. With satellite imagery, an analyst can differentiate varieties of vegetation and monitor their health, thanks to three characteristics of plants, chlorophyll, leaf structure, and water content. Obviously the last one, water content, is critical to determining the fire threat.

The more water a leaf contains, the less mid-infrared light is reflected into space. Obviously, this effect provides an ideal tool for measuring the impact of drought on a forest. Information like this is invaluable to forestry managers who want to know where vegetation is dry enough to fuel a fire.

The traditional tool for studying a forest is the color infrared image. This image highlights all the vegetation in a scene and gives the viewer an idea of the condition of the plants. Sick plants will either be a pale green or red (depending on the type of color infrared image), while healthy plants will be a bright color.

Sunken Spanish Galleons and Buried Pirate Treasure (Archeology). Satellite imagery has also been used to hunt for Spanish galleons in the Caribbean, off the coast of Belize. The coast is protected by barrier reefs upon which old ships often ran aground, before better navigation techniques were discovered. Since the visible (blue, green, and red) bands can be used to see underwater, scientists have used satellite imagery to see anomalies on the ocean bottom. The idea is that a sunken ship will be higher than the surrounding bottom and will reflect light differently. Although the anomalies might not be ships, it narrows down the area to be searched. So far this technique hasn't found anything rewarding, but it remains a valid technique for the treasurer hunter.

There are other potential treasure hunting methods that can use satellite imagery. Scientists also looked at the coast of Nova Scotia, where pirates often came to bury their treasure. The idea was to look for unusual features along the shoreline, but nothing came of it. “There were too many unusual features,” said one imagery analyst involved in the project. However, if this technique were combined with some good data from other sources, well, who knows!

There are others who want to use satellite imagery to find caves, where treasures may be hidden. Once again, nothing of value has been found using this technique, but imagery can be used to narrow down the search area by locating limestone formations, where caves are common.

Monitoring the Colorado River (Water Consumption). No matter how low the temperatures are in the East and Midwest, consumers know they can walk into a grocery store and find fresh fruits and vegetables. However, these out of season fruits and vegetables can only grow if they are watered by the Colorado River, and unrestrained demand would turn it into a muddy ditch if it were not constantly monitored. While the East and Midwest can be over-endowed with snow and ice during the winter, water is still a precious commodity in the Southwest.

Although pumping stations along the river measure the amount of water taken from the river, it is not a true picture of actual demand. Much of the water is used for agricultural purposes, and a large percentage of the water used for irrigation returns to the river via the ground water table. The only water that is actually consumed is that which evaporates through the leaves of plants, a process called evapotranspiration. Naturally, this percentage varies depending on the type of plant under irrigation, the temperature, and the hours of sunshine.

Determining agricultural water demand was once labor intensive. It required taking agricultural reports of what was being grown and determining water consumption by multiplying a crop’s evapotranspiration rate by the number of acres.

Since the near infrared band shows plant health, it visibly shows if the field is being irrigated or the plants are in need of more water. Although there are differences in reflection depending on the type of plants and even if there is dust on the leaves, this drastically reduces the workload for the Department of the Interior.

Open Skies. (Reporting the news). The Iraqi war of 2003 was a landmark in the use of satellite imagery in reporting the news. Nearly every news organization had imagery of Iraqi targets, which was being updated on a regular basis. Viewers had an unprecedented view of what was happening behind Iraqi lines.

Despite this success, news organizations were slow to realize the uses of satellite imagery. The first application was in 1986, when the nuclear reactor at Chernobyl exploded. Given the secrecy of the Soviet government at the time, it was up to the LANDSAT and SPOT satellites to show and prove that there was actually an accident deep in the Ukraine. Information from LANDSAT 5 showed a bright area that was too strong to be a reflection of the sun. The only answer was an energy source with a temperature of about 1,000 degrees Celsius, meaning that the fire was still raging in the reactor's graphite core. Data from the satellite's thermal sensor showed that the cooling pond had a uniform temperature and wasn't being used for cooling, which meant that the reactors had been closed down for an emergency.

France's SPOT satellite soon confirmed the LANDSAT information and showed even more with its higher resolution panchromatic images. SPOT not only showed the damaged plant, but also even showed the scars of the blast. Thanks to the public revelations from the two satellites, the Soviet government was forced to be more open about the incident. And, before you knew it, news agencies were scouring the world to find news stories that were once unobtainable to them.

The first story to use satellite imagery in investigative reporting was the ABC News report on the Soviet Krasnoyarsk radar facility. For years the U.S. government had accused the Soviets of violating the ABM treaty by building the Krasnoyarsk facility to track American ICBMs in their terminal phase. The Soviets vehemently denied it and claimed that the facility was for tracking its own spacecraft. The situation reached an impasse since many Reagan Administration critics felt the accusation was an attempt to push for more defense spending.

On April 2, 1987, ABC News used SPOT imagery to prove that the position and orientation of the radar made it impossible to track Soviet spacecraft launchings and the facility was, in fact, a violation of the ABM treaty. One result of this disclosure was that the Soviet government eventually agreed to dismantle the radar.

Since then, several news agencies have used satellite imagery in investigative reporting. CBN (Christian Broadcasting Network) News has been a pioneer in using satellite imagery in following the growth of weapons of mass destruction in the Middle East and North Korea. Often, their reports have elicited calls from congressional intelligence committees that have asked for more information because the investigative report provided more information than U.S. intelligence groups were willing to disclose.

As spatial resolution has improved, the amount of reporting on military issues has increased. Today's satellite imagery can clearly see and identify tanks, aircraft, and missiles. It can also peer into secret areas once closed to the public. And, with multispectral data, reporters can even detect camouflage and heat sources from underground facilities. This is one part of the satellite industry that will grow in the future.

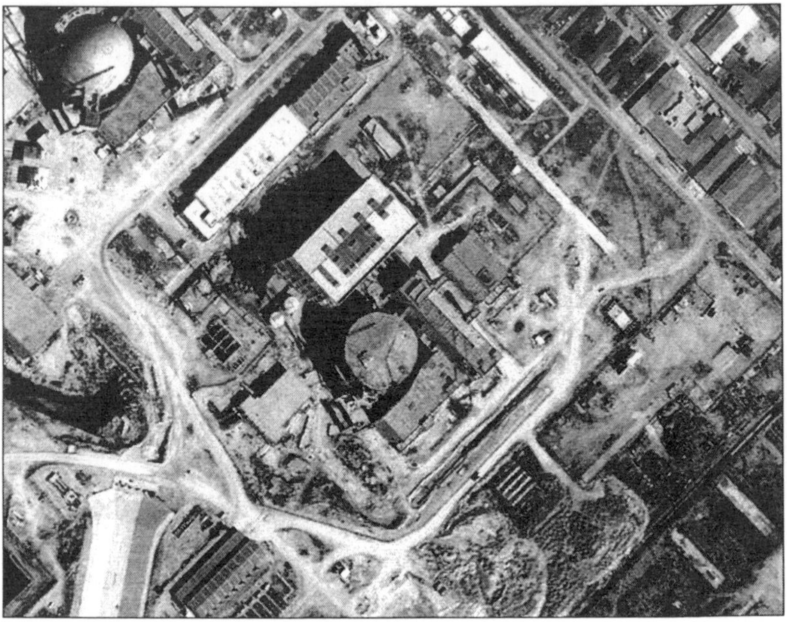

Figure 4.1

News groups that follow the proliferation of nuclear weapons have constantly monitored Iran's nuclear reactor at Bushehr. The reactor containment vessel in the center of the picture is still under construction. The one in the upper left hand corner is finished. (Space Imaging and CBN News)

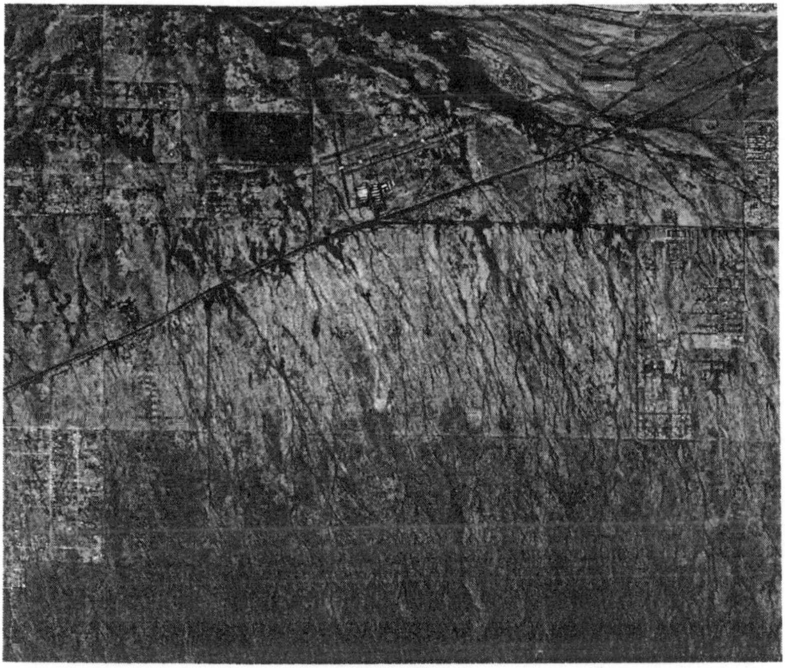

Figure 4.2

As areas like the west side of Tucson, Arizona, rapidly grow, satellite imagery is often the only way to keep track of development. I also use this type of imagery to find the best places to hunt for morning dove (dark areas where there is vegetation). (Space Imaging)

Down on the Farm (Agricultural Studies). It's part of the American farming tradition and as familiar as a Norman Rockwell print; a grizzled farmer steps outside his house, looks into the sky, and makes some quaint remark like, "we sure could use some more rain." Of course, American farmers are not the only ones who do this. For thousands of years, all farmers have looked to the sky for help, whether it be for rain or sunshine. Today, however, farmers are looking to the sky for up-to-date satellite information.

Although satellite imagery has always held promise for farmers, it wasn't until recently that the imagery could be acquired and delivered fast enough that any potential problems could be rectified. For instance, a pest or plant disease that was starting on a field had to be detected and countered within a couple of days, or it would be too late to make a difference. That was hard to do in the past, when delivery times were longer and satellite costs were higher.

Today, a satellite can image an area, the data can be down linked to Earth, it can be processed, and it can be electronically transmitted to a farmer within the 48 hours of acquisition necessary to fix the problem. One of the common problems is plugged or improperly sized sprinklers in an irrigation system. Since the amount of water can dramatically affect a crop yield, the faster the information is available, the faster the sprinkler is fixed.

Selling the Public (Public Relations). Placating some environmentalists was actually easier than moving an eight foot square satellite image around Jacksonville, Florida. Consultants had to manhandle it around corridors, into elevators, and through rooms in order to make presentations. They even had to build a special carrier for the top of a car to carry it across town.

Difficult as it was to move this clumsy display around, the "grunts" charged with moving it agreed that the image of 730 square miles of northwest Duval County was indispensable to successful public meetings and presentations concerning Jacksonville's outer beltway. "It was a great presentation display," one manager noted. Without it, Jacksonville roads would have become as crowded as the elevator carrying the 8′ x 8′ image.

The Florida Department of Transportation was trying to sell an outer beltway for Jacksonville, but it was meeting with opposition from environmentalists afraid that the road would destroy wetlands, and by those who oppose growth of any sort. Although the original studies took the environment into consideration, the Department had to convince the opposition.

The large satellite image was a bother to move, but it was critical for the campaign. The image added credibility to the analysis at public meetings. By bringing the large photo, which showed wetlands, residential areas, and flood plains, the civil engineers took skeptical observers though the decision making process and showed what steps they took to limit environmental impact. As a result, opposition never coalesced as it has with other civil engineering projects.

One Florida engineer noted that most civil engineering presentations are poor and a satellite image helps the public relate better to the project. By showing environmentally sensitive areas, waste dumps, and residential areas, it reinforces why the decisions were made.

Witness for the Prosecution (Legal Evidence). How do you prove that a landowner has changed the ecosystem after a law was passed? That was the problem for the South Florida Water Management District, which is charged with monitoring over 100,000 acres of wetlands where farming takes place. The area is too large for traditional surveillance methods and much of it is far away from roads.

When the district finds evidence of environmental tampering, they have to prove that the work took place after wetland protection laws were passed. The answer was archival satellite imagery and newly acquired images. By electronically comparing the new and old imagery, helicopter pilots could be given locations of possible violations. If a change was detected, the old satellite imagery can prove that the changes were made after the law was passed. Usually, when presented with this evidence, the land owner will chose to settle out of court and pay a fine.

Of course, what's sauce for the goose is sauce for the gander. Someone accused of damaging the environment can use old satellite imagery to prove that the damage occurred either before the law was passed or before the property was transferred.

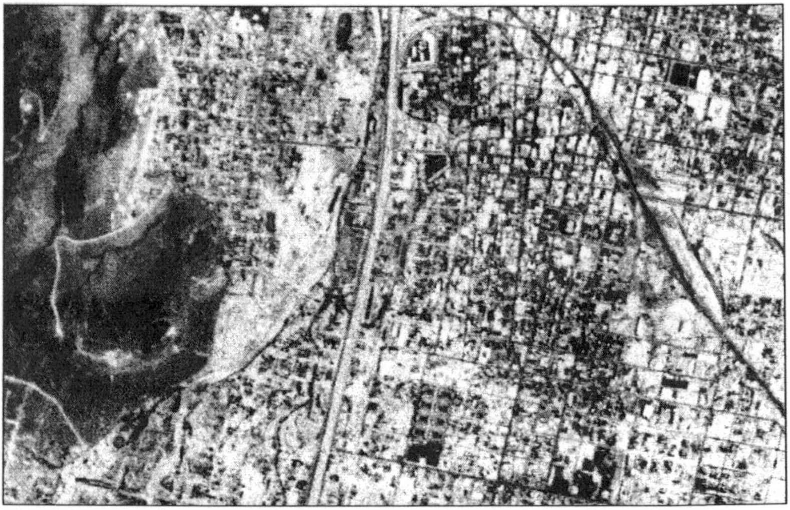

Figure 4.3

Downtown Tucson as it was in 1965. The historical barrio has been replaced by modern buildings for bureaucrats and lawyers.

Zoning, Land Use, and Permitting (Taxes and Law Enforcement). Unfortunately, Big Brother is here and he is watching you from space.

Since it's nearly impossible to send inspectors everywhere, more and more local governments are relying on satellite imagery to detect changes that would indicate construction or some other modification. Any changes can then be checked against tax and permit records, and inspectors can be dispatched as necessary. The result is more tax money for the government and more government control over your life.

Of course, satellites aren't perfect and it's possible that they can make a mistake. In a later chapter, we will cover how satellites detect changes in land use applications and their weaknesses.

The Way We Were (History). Satellite imagery is often a window into the past. In this case, it is the rapidly growing city of Tucson, Arizona. Figure 4.3 is the first known satellite image taken of Tucson, Arizona, by Mission 1021-1 in May 1965. The image shows the historical barrio that was torn down by city leaders in the name of urban renewal in the 1970s. With imagery like this and software that can make a satellite image a three-dimensional image, historians can recreate a scene that has been destroyed forever.

Chapter Five
A Thumbnail Guide to Interpreting

Needless to say, this is a subject that whole books have been written about, so there is no way that I can cover it with the detail that you may want. However, I'll try to cover the basics and then go into more detail on several areas that may interest you.

Basic Principles of Object Recognition

Thanks to modern air travel, most people are familiar with viewing objects from overhead and, therefore have little difficulty recognizing many features like roads, lakes, residential areas, etc. However, in order to become an accomplished satellite imagery interpreter, you need to engage your mind as you look at the imagery. Here are some of the factors that you should evaluate as you look at any satellite image.

Shape. This is probably the most critical, because shape alone will identify many objects like airports, roads, etc.

Size. Size is critical, especially if you are differentiating between a rural road and an interstate highway. In most digital satellite imagery, it's relatively easy to determine size and distance because the spatial resolution of the image is usually the size of one pixel of the image. Therefore, if the resolution is one meter and the object is four by three pixels in size, the object is about four by three meters.

Form Follows Function. Most objects are built in a certain way for a reason. In military reservations, wide curving roads are often built to allow the movement of long vehicles like missile transporters.

Tone and Color. If the image is a high-resolution panchromatic image, the different colors will register in different tones and shades. This can mean the difference between a tile and a slate roof or a cement or asphalt road.

If you have a color image, basic colors can tell a lot. For instance, a rectangle in a back yard in the suburbs that is blue, is probably a swimming pool. If the shade of the blue goes from light to dark blue, the light shade is the shallow side, while the dark is deep.

Time. Be aware of the day and time of day the image was taken. A parking lot that is full on a weekday may be a place of business, while one full on the weekend may be a shopping or recreational center.

Several years ago, a major international defense publication asked a photo interpreter to look at the Island of Abu Musa, which belongs to the United Arab Emirates, but was occupied by Iran several years ago. Since then the Iranians had fortified the island. However, the analyst reported that Iran wasn't involved in building any additional facilities, because he didn't see any construction activity. If he had paid attention to the date the satellite image was acquired, he would have known that it was Friday, which is holy day for Muslims. A picture taken any other day of the week would have shown considerable activity.

Pattern. Patterns can indicate objects like parking lots. In agriculture, if trees are arranged in rows, it's obvious that they are a cultivated orchard rather than wild vegetation. If a field is plowed in a curve, rather than a straight line, it probably indicates a slope.

Site. The location of an object is helpful in identifying an object. A silo would be a logical identification of a tall cylindrical object sitting next to a barn. It might be a bit harder to identify if it is sitting in a mall parking lot (it might be an inflatable Budweiser beer can).

Shadow. This is critical for measuring heights and for recognizing tall objects like smokestacks and antennae. Unfortunately, they can also hide detail next to buildings. However, if you refer to the book *Satellite Surveillance*, you can see how a computer can reveal details hidden in the shadow.

Drainage. Looking at drainage, creeks, and rivers can give you an idea of relative elevation, soil conditions, and even vegetation. A wandering river or stream means that the terrain is flat.

Texture. The amount of coarseness on the object can indicate what it is. For instance, it is often hard to tell if a military vehicle is wheeled like an armored car, or tracked like a tank from space. However, by looking at the ground around the vehicle, you should see the scarring caused by treads, especially in turns.

As you start interpreting the image, begin to look for interrelationships. If the image is of a rural farming community, you should see barns, silos, pastures, livestock and crops. There may likely be some signs of food distribution (depots, trucks, warehouses, train system, etc.). From these observations, you can make some general assumptions about the area and people.

Since it can be easy to focus on one factor, photo interpreters often use checklists to remind them about what they should be focusing on. Although we will be going into more detail in later sections, here's a checklist of basic features that you will want to make a note of in a basic interpretation.

Natural Features

Forests
Tree farms
Natural vegetation
Pastureland
Crops and fallow fields
Irrigated crops
Orchards
Parks
Shorelines and beaches
Livestock
Bays and inlets
Swamps or marshes
Floodplains
Rivers and streams
Lakes and ponds
Sand bars and mud flats
Limestone and sinkholes
Cliffs
Erosion features
Rock outcrops
Water drainage
Faults
Talus slopes and alluvial fans
Hogbacks

Manmade Features

Apartment buildings
Mobile homes
Private homes
Schools
Stores, shopping centers, and malls
Government buildings
Churches
Parking lots
Sports fields
Secured facilities (prisons or military compounds)
Open pit mining
Quarries and sand pits
Underground mining
Oil and natural gas wells
Electrical power plants
Electrical power distribution systems
Buried pipelines
Petroleum and Chemical industries
Manufacturing

Natural Features	Manmade Features
Anticlines	Roads (dirt, two lane, four lane, etc.)
	Railroad lines, yards, and terminals
	Truck terminals and warehouses
	Airports
	Radio or TV towers
	Dams
	Bridges
	Ports and ferry landings

Developing an Expertise

One of the nice things about interpreting satellite imagery is that it is so democratic. Everyone has a specialty that sets him or her apart from the rest. For instance, a truck driver could look at an image of a truck depot and warehouse and tell us more about the operation than a professor in photo interpretation. Landscapers, longshoremen, farmers, military people, and any other occupation have a unique perspective that they can bring to satellite photo interpretation. Therefore, it pays to specialize in what you know. Conversely, if you have lived your whole life in New York City and want to study agricultural imagery, in order to make a killing on potato futures, you are probably going to blow it.

Of course, you probably won't know enough to be an instant expert, but if you go with your profession or interests, you have a head start. The next step is to expand your knowledge based on your area of expertise. For instance, if you are a landscaper or enjoy gardening, you probably have a better understanding of vegetation and its requirements. Build on that by studying more about forestry and agriculture. As you go along, you will find areas that you understand better. Continue to focus on those areas.

As you improve, you will want to buy more books that will help you in satellite interpretation. These are often available at used bookstores near universities that offer aerial and satellite interpretation courses. Don't be put off if you find an old book about aerial interpretation. Given the current spatial resolution of satellite images, they are very similar to aerial photographs.

There are other types of books that can help as you develop your specialty. College textbooks on industrial processes and engineering can help you identify certain industrial sites. A college chemical engineering book can help identify chemical processes, even if you can't understand all the writing. And military history books can often give insight into what is happening on a military base.

Of course, I can hear you asking, "How can I, an amateur, hope to compete with professionals?" Actually, it is easy. Right now, there is more satellite imagery than there are satellite analysts. That means that there are a lot of things left undiscovered and underutilized in satellite imagery. Although you may not be as good, if you spend more time on one satellite image, you will see things the professionals missed because they were looking at it too quickly. Besides, you may have other advantages like a greater familiarity with the area, the chance to go back to the image often, and a personal passion that the professional doesn't always have.

So much for the basics. Let's look at some specific areas of satellite interpretation that can benefit you.

Archaeology

Aerial imagery has been used in archaeology since the 1920s. Its first application was accidental and occurred when some Royal Air Force photographs were shown to Dr. Crawfors, who founded the magazine *Antiquity*. These photos clearly showed the pattern of a Celtic field system from overhead. Since then, it has been a critical part of uncovering lost sites and, as a result, in World War II many archeologists in Britain were employed in military photo interpretation. Today, as we noted earlier in the discovery of the ancient city of Ubar, satellite imagery is beginning to fill the same role.

There is probably no better example of using satellite imagery in the United States to find an undiscovered archaeological site than Chaco Canyon, New Mexico. Although they could not be seen from the ground or even the air, infrared sensors detected prehistoric Indian roads in the canyon that date from 900 or 1,000 A.D. These roads, which were 20 feet wide, were seen as straight lines on the satellite image. In later research, the scientists discovered over 200 miles of roads that apparently connected the people of the area to sacred places and Chaco Canyon.

One of the most important indicators of archaeological sites are "plant marks" — visible differences between vegetation on a site and that in the surrounding area. For instance, in an area where there is an ancient ditch that has since been covered up by dirt and humus, vegetation will often be more lush because water seeps into the ditch and accumulates there, especially during dry times of the year. Conversely, a stone wall, road, or foundation that has been covered up will inhibit root growth and interfere with moisture travel, which will stunt plant growth.

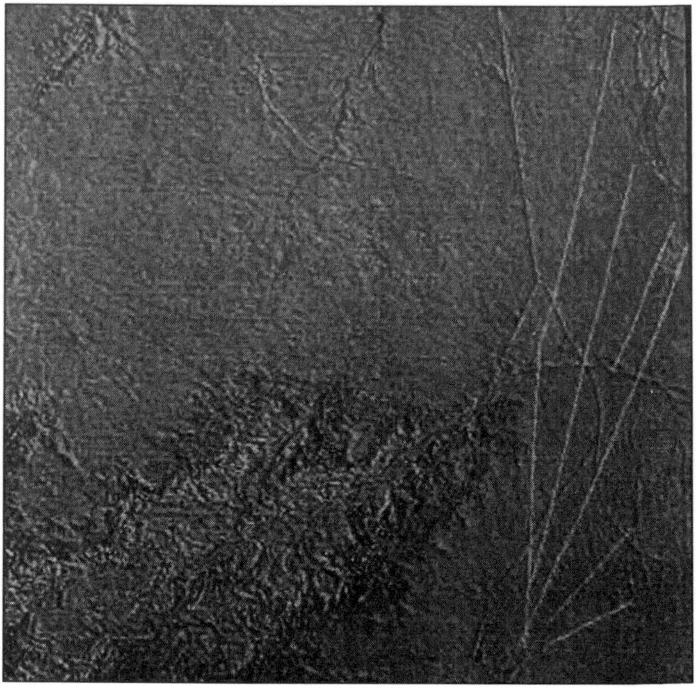

Figure 5.1

This imagery of Chaco Canyon, New Mexico, shows several straight Indian roads (right hand side of the image) that were built over a thousand years ago. (NASA)

Grasses seem to be the best indicator of uncovered walls or ditches since the plants are sensitive to variations of the soil and grow closely together. In the Great Plains, it's not uncommon to find circles that indicate where Indian lodges were built centuries ago. Of course, some of those circles are nothing more than buffalo wallows, so ground research is necessary.

The best time to look for plant marks is during the growing season and the best satellite imagery to use will include the near infrared band, which highlights plant health. If possible, the image should be acquired after a short dry spell because the vegetation will be more stressed and any differences in plant health will be more pronounced.

The same principle can also be used in dry areas like the Southwest, where it may take centuries for vegetation to recover from damage. In arid parts of the Southwest, once the vegetation has been destroyed it's very hard for the soil to recover. The bacteria and nutrients common in wetter climates are missing, which makes it harder for vegetation to take root again. In time, rainfall erodes the little useful soil, which makes recovery that much harder.

Thanks to the fragile Southwestern environment, it's relatively easy to discover signs of the Western movement. Some of the objects that can be seen are trails, mining camps (often very visible thanks to the mine tailings and excavation), forgotten ghost towns, and abandoned homesteads. Unlike the grasslands of the Great Plains, infrared imagery doesn't have the strong infrared reflection from the sparse vegetation. However, barren land usually is quite bright, so the high resolution panchromatic imagery is often good.

This brings us to the next indicator of archaeological sites, soil marks. In the Eastern United States, Indians who lived near the ocean would often eat clams and mollusks. Consequently, scattered seashells often mark their sites. In areas where there is less vegetation, Indian sites can be detected by looking at pebbly areas, where rocks have been cleared.

Trails are often also found by soil marks. Whether it is the hooves of animals or wagon wheels, in the Southwest, the dry soil would be broken up and blown away as dust, leaving brighter rocks exposed.

Old mining sites are a natural for soil marks. Tailings piles of waste rock or processed ore will likely be a different color or tone (in a panchromatic scene) than the surrounding soil.

In any of these cases, a high resolution panchromatic image is one of the best tools for soil markings. If you want to use multispectral imagery, the mid infrared bands (1.55 to 2.35 microns) will show different soils and their moisture levels quite well.

Shadow marks are also a useful tool for archaeologists, although most satellites are programmed to fly over a site in the mid morning, when shadows are less. The theory is that walls and other high points underground will throw a shadow. This works best in the early mornings and late afternoons, but satellite images taken in the mid morning may still help in discovering an archaeological site. Again, high resolution panchromatic images are the best for this type of research.

Another technique for finding archaeological sites is to look for habitability marks. For Indians, habitability marks might include access to fresh water and lush vegetation, where animals might be found (hunter Indians) or berries and fruits are common (gatherers). In this case, mid infrared imagery will highlight healthy vegetation. In areas of the world where water might be scarce, satellite imagery might be used to find signs of ancient catchment pools where water would be stored.

Military Interpretation

Although satellite imagery is more and more a civilian industry, military affairs still cast a large shadow over the business. In this case, there are two aspects: military use of civilian imagery and civilian monitoring of military issues.

Although the military and intelligence communities have a fleet of spy satellites, they still use commercial satellites for intelligence purposes, as evidenced by the inability of commercial clients to acquire satellite imagery over the Middle East during the Iraqi Campaign of 2003. The fact is that the U.S. intelligence community has tens of thousands of targets that it wants to monitor and it's hard to monitor all of them with its own constellation of satellites.

Although commercial resolution isn't as good as that of the spy satellites, there are still many missions that they can handle. Coarser spatial resolution imagery like that from LANDSAT can be used for crop harvest estimates, maps, cross-country mobility, and construction. Higher spatial resolution imagery like Ikonos and Quickbird can be used to identify and count equipment, develop targeting data, and monitor lower priority installations. It can also be plugged into training devices so pilots can practice bombing missions.

Satellite imagery can also be used by the military in many cases where spy satellite data is too sensitive to use. Reconnaissance patrols can use it to have a current source of data for the area they will patrol. Snipers can use imagery to plan infiltration routes, calculate fields of fire and determine distances before a mission. Patrols can use imagery to see if a trail has been recently mined by looking for the scarring that indicates disturbed soil. Ambushes can be planned in advance. Mission information can be given to everyone involved without worrying about compromising intelligence sources. The applications are limitless.

One of the most popular uses of commercial satellite imagery is following military developments that were once hidden from the general population. Over the past ten years, satellite imagery has been used more often to show television viewers foreign military installations that pose a threat to world security, especially those that either produce or store weapons of mass destruction. In that regard, they have made for a more informed electorate, but at the same time, have been used by some groups to push their own agenda.

One example of agenda driven satellite imagery interpretation was North Korea's missile test site at Taepo Dong. One Washington-based think tank that monitored WMD development, but opposed the development of an American anti-ballistic missile program, released imagery of Taepo Dong and claimed that its interpretation showed the site was too simple and crude to allow for development of a North Korean missile that could reach the United States. Therefore, they reasoned, there was no reason for developing an ABM system.

However, other analysts looked at the imagery and saw enough of an infrastructure to warrant concern by America. Inevitably, that concern was borne out, as it was later revealed that these nuclear missiles could reach the U.S. All the imagery proved was that a simple site could be used to develop long range ballistic missiles. You don't need a Cape Canaveral to produce and test ICBMs.

The growth of this civilian defense monitoring industry has created a need for more information on what to look for. With that in mind, here are many of the military objects that can be found at military installations.

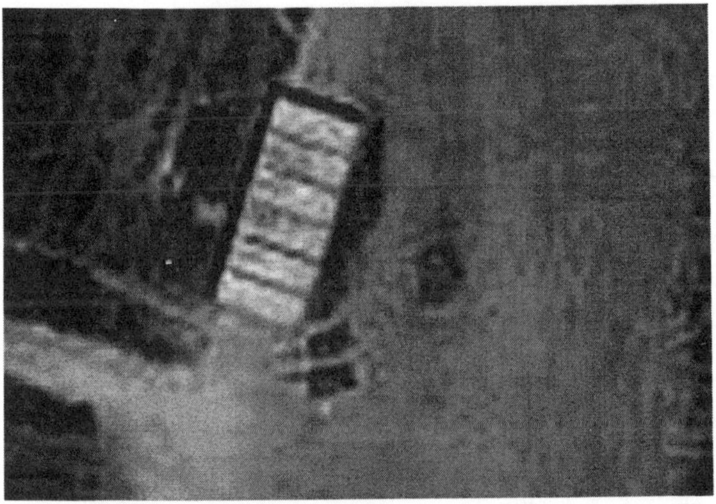

Figure 5.2

Thanks to higher resolution satellite imagery like this one meter Ikonos image; one can see the barrels that identify these vehicles as Syrian tanks. (Space Imaging)

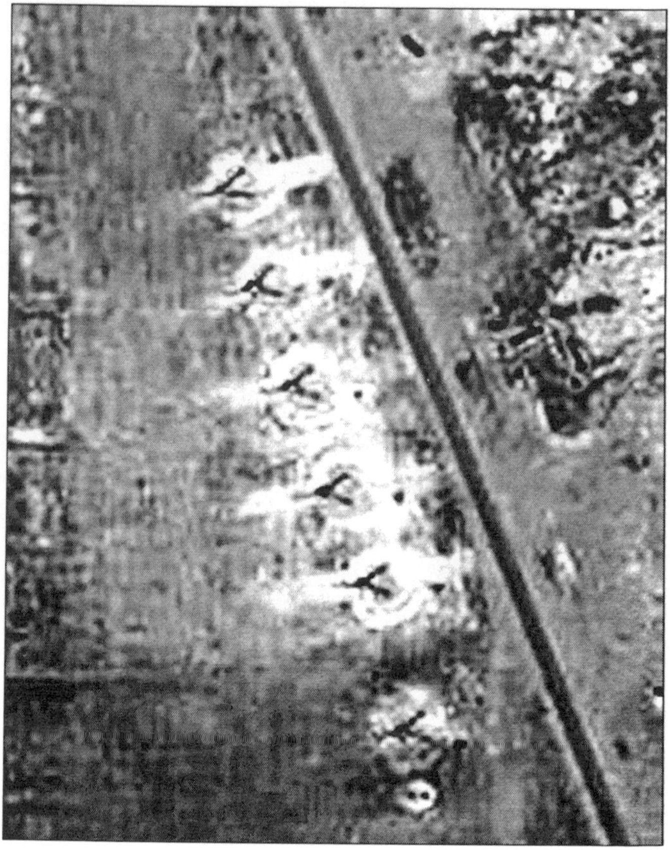

Figure 5.3
These are Syrian howitzers positioned in revetments. (Space Imaging)

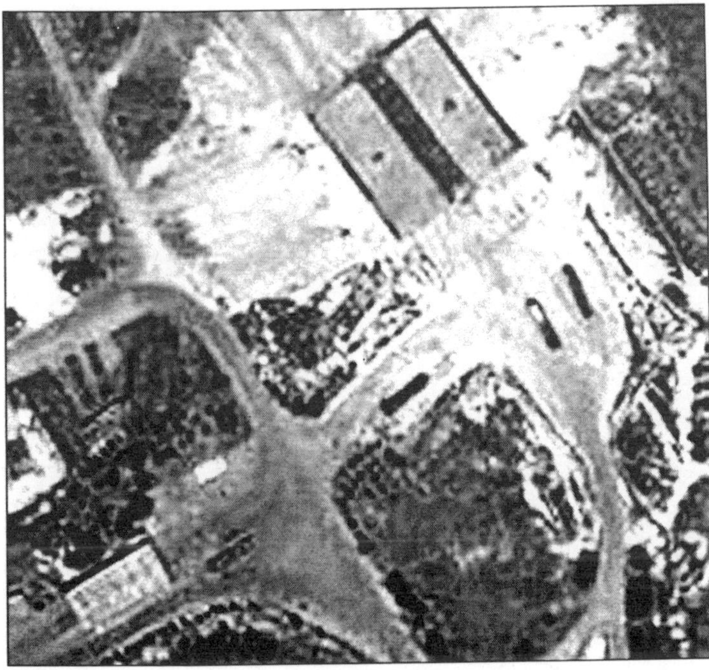

Figure 5.4

Several Syrian Scud transporters are seen outside buildings at this facility near Homs, Syria. The building in the lower left hand corner is a drive-through building that is used to accomplish maintenance on missiles or to load warheads. This is a critical feature at missile sites. (Space Imaging)

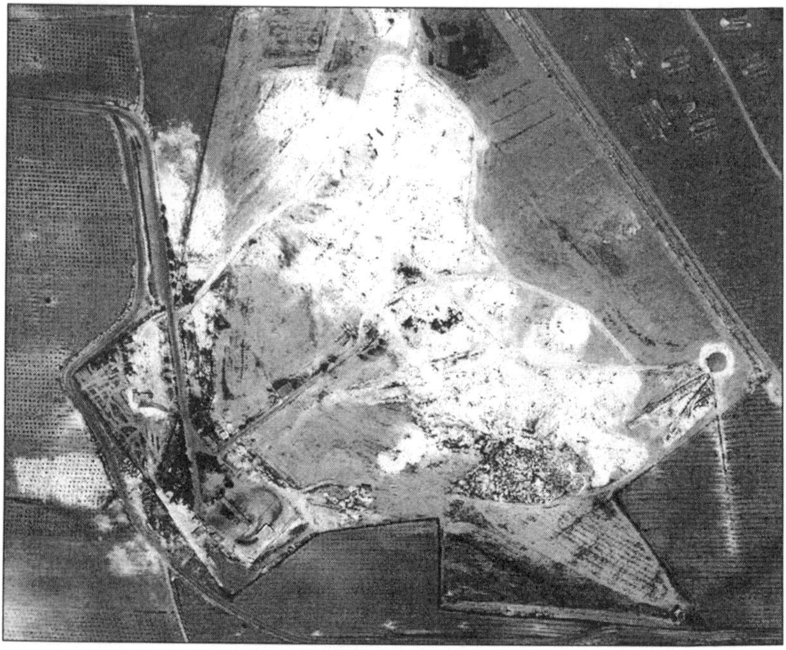

Figure 5.5

This suspected chemical weapons facility near Homs, Syria has several interesting features. It has a clear fence perimeter surrounding the facility, which indicates its sensitivity. It has a gentle curving road coming in from the upper left side of the image, which indicates it is probably used by missile transporters, which can't make sharp turns. It has a drive through building in the lower left hand side, where missile warheads can be mated to the missile. There are also several cement igloos inside the perimeter (one is clearly visible on the right side of the image). (Space Imaging)

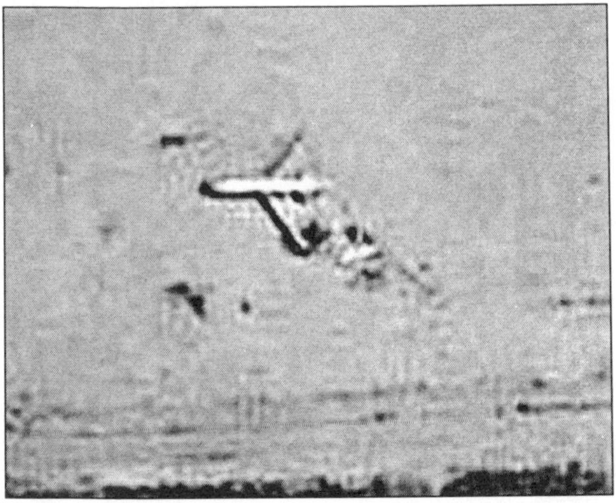

Figure 5.6

Look for the unusual. Large jet aircraft aren't unusual at airports, but they are rarely found on an island without an airport. This passenger aircraft is at the terrorists training facility at Salman Pak in Iraq. Terrorists could sharpen their hijacking skills here before taking a civilian aircraft with hostages. (Space Imaging)

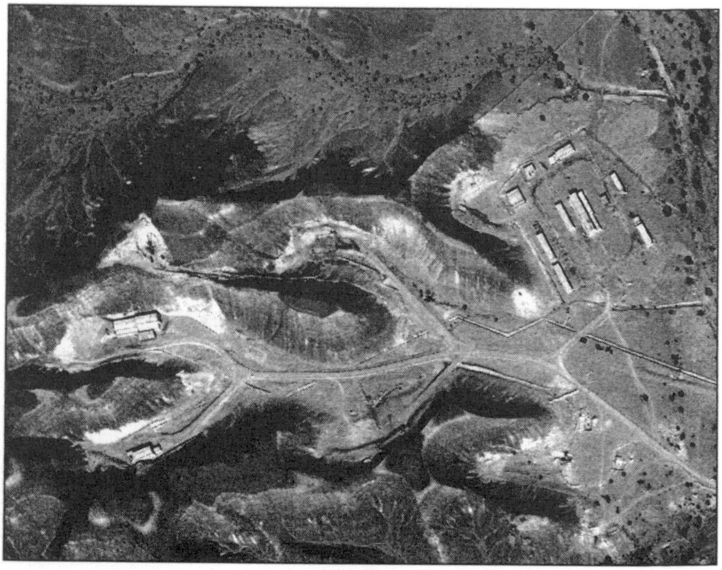

Figure 5.7

A ballistic missile site in Saudi Arabia. On the left hand side of the image are several bunkers, where the missiles are stored. On the right hand side is a missile maintenance site. Note the drive-through buildings for the missiles and their transporters. (Space Imaging and CBN News)

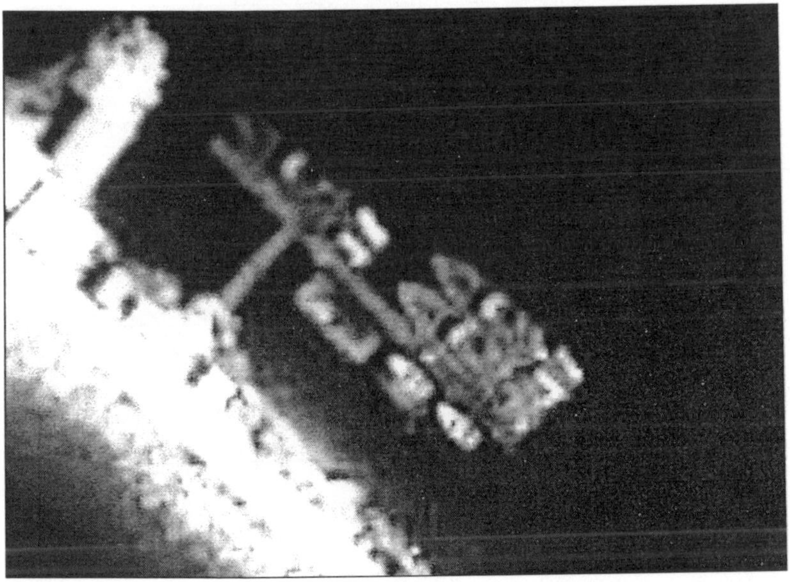

Fig 5.8

Three Chinese-made Hudong missile boats docked next to the Iranian nuclear reactors at Bushehr. With one-meter resolution, and reference books like Jane's Fighting Ships, one can identify most military naval vessels. (Space Imaging and CBN News)

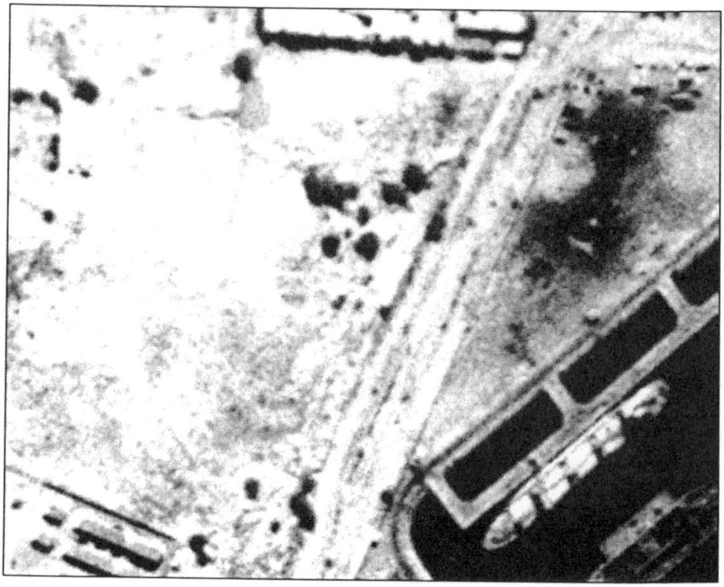

Figure 5.9

Merchant ships can be identified with Jane's Merchant Ships. The ship seen in this Russian KVR-1000 satellite image is a Chinese ship delivering missiles to the Iranians in Bandar Abbas. The missile containers can be seen on the deck of the ship. (Sovinformsputnik)

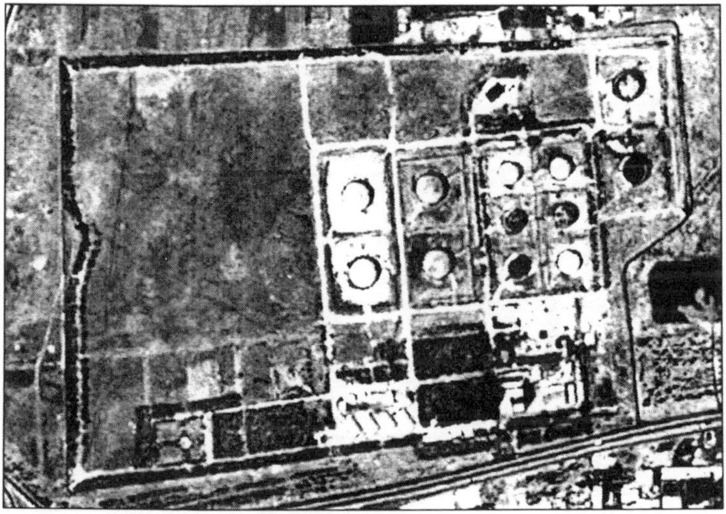

Figure 5.10

Sites involved in chemical weapons production are often found near refineries. This suspected chemical weapons precursor production site is near Homs, Syria. Unlike the nearby refinery, this site is surrounded by two fences and guard towers (always check places with security perimeters). (Sovinformsputnik)

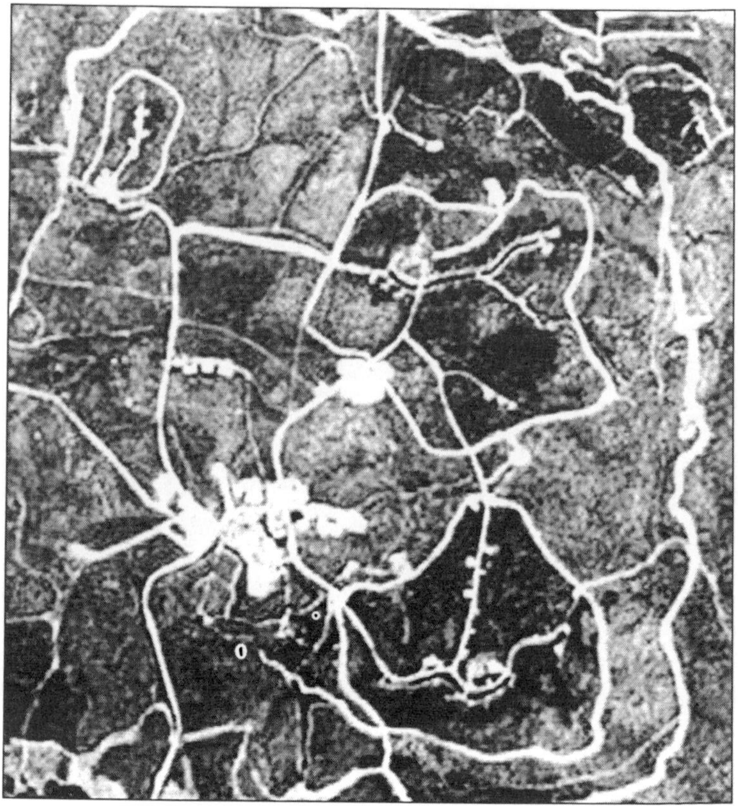

Figure 5.11

Even lower resolution satellite imagery has uses. This Russian image of Israel's nuclear missile site at Zackaraya has a spatial resolution of 5 meters, but still clearly shows the perimeter and roads leading into bunkers in the side of hills. (Sovinformsputnik)

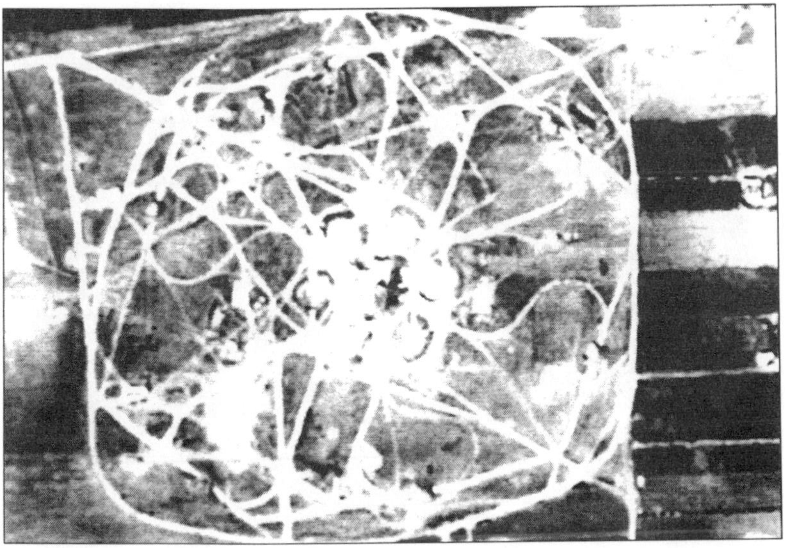

Figure 5.12

This is a classic Russian surface-to-air missile site taken by a Russian KVR-1000 satellite. The radar is in the center, while the missiles are on roads on the outer perimeter. (Sovinformsputnik)

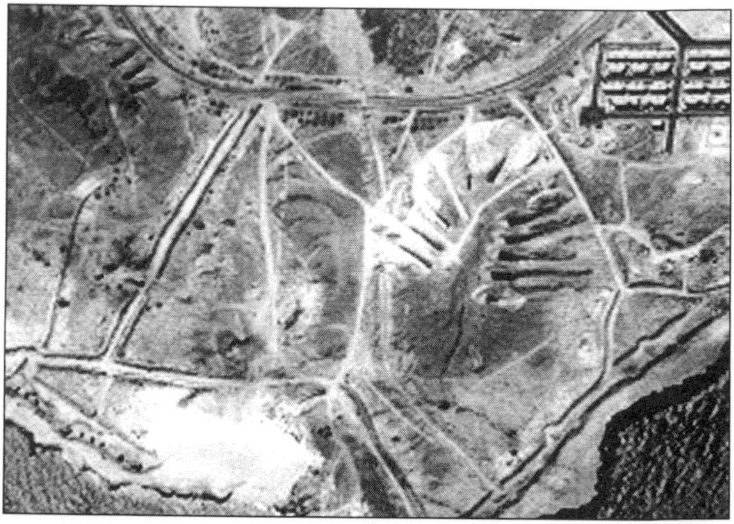

Figure 5.13

Silkworm anti-ship missile bunkers on the island of Abu Musa. Missile bunkers are often dug into the side of hills, where they can be kept until moved to their launch pad in the lower left hand corner of the image. Military barracks are visible in the upper right hand corner. This island, owned by the United Arab Emirates, but illegally occupied by the Iranians, sits astride the world's oil shipping lanes. These missiles pose a threat to every tanker that goes into the Arabian Gulf. (Space Imaging)

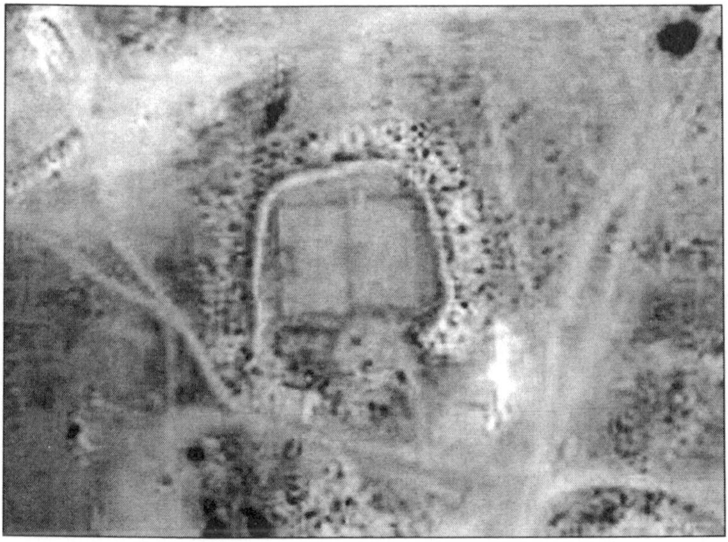

Figure 5.14

The open area inside the revetment is probably a pre-surveyed launch site for an Iranian surface-to-surface ballistic missile. In order to limit the time the missile transporters are out in the open and vulnerable, engineers clear and survey the site. The missile operators already have the targeting information, so they merely have to raise the missile and launch it. As you can see from the faint markings, the site has been camouflaged as a soccer field. (Space Imaging)

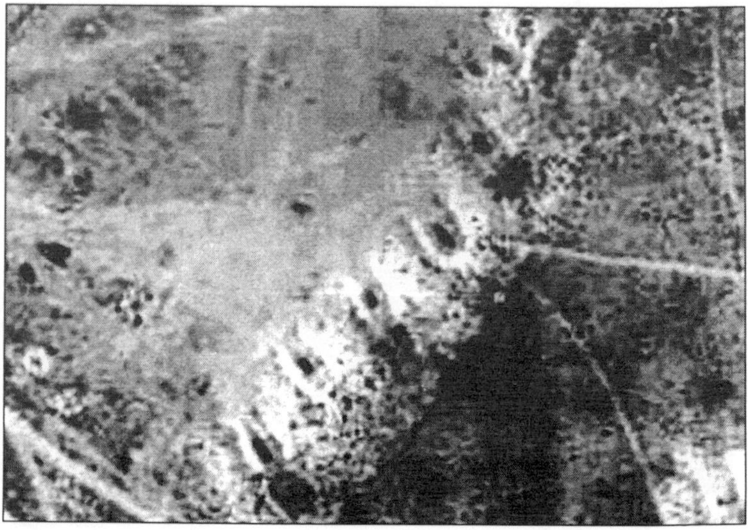

Figure 5.15

Silkworm missiles on the southern side of Abu Musa. While many analysts may think they are tanks, it is obvious that one meter resolution imagery can show the gun barrel (see the earlier image of Syrian tanks), which these lack. Note that the area behind the vehicles isn't scarred from the tracks, so they are wheeled. (Space Imaging)

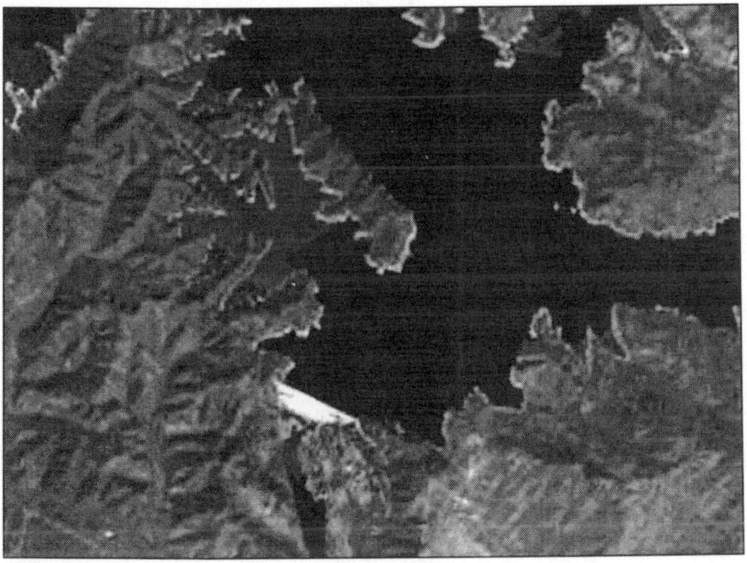

Figure 5.16

Dams and the electricity that they produce are often critical to weapons of mass destruction production. This North Korean dam (lower center) is critical to nuclear weapons development. The satellite image is a Russian KFA-1000 with five-meter spatial resolution. (Sovinformsputnik)

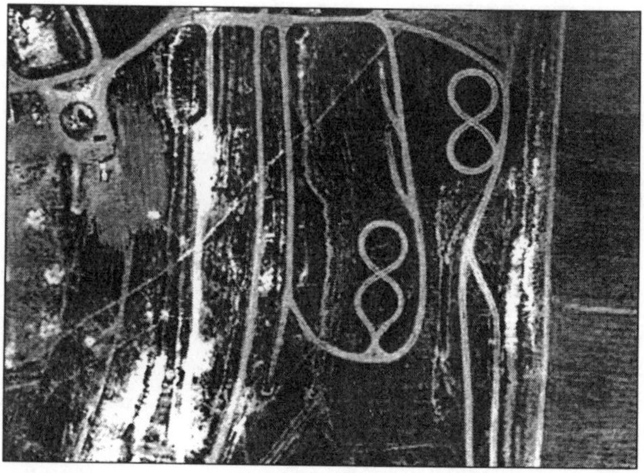

Figure 5.17

Training sites for missile transporter drivers have driving courses where they can practice sharp turns. Note the figure eights, which are a regular feature. (Space Imaging)

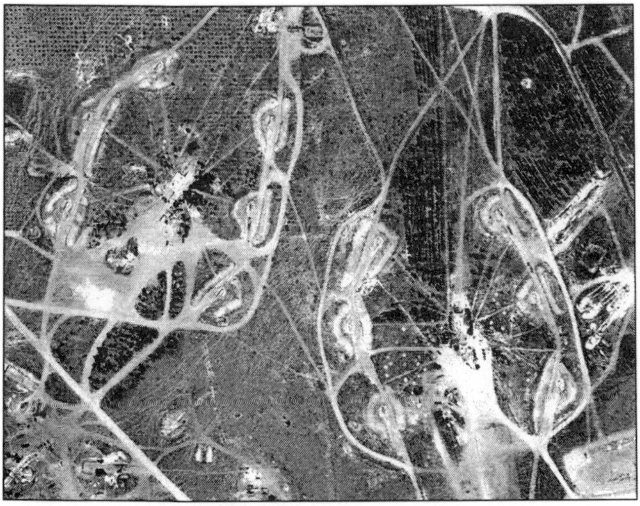

Figure 5.18

Syrian missile launch pads. This site near Homs, Syria is used by Syria to test its Scud missiles. Revetments surround these two launch pads where vehicles and testing equipment can be parked during a launch. (Space Imaging)

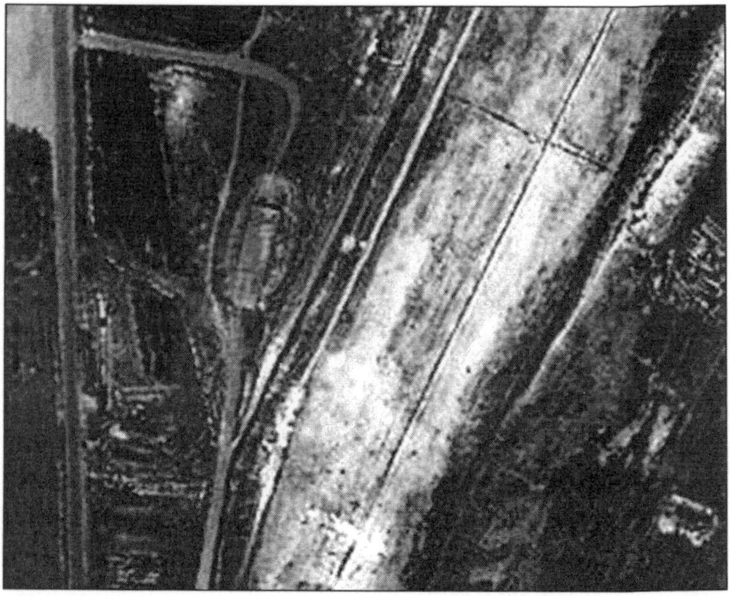

Figure 5.19

Another drive-through building (left center) and a gentle curving road (left upper corner). The empty decontamination pool next to the building (right side) means the building is probably used for loading toxic fuels onto Syrian missiles. The rows on the left side are trees that obscure the building from the road. (Space Imaging)

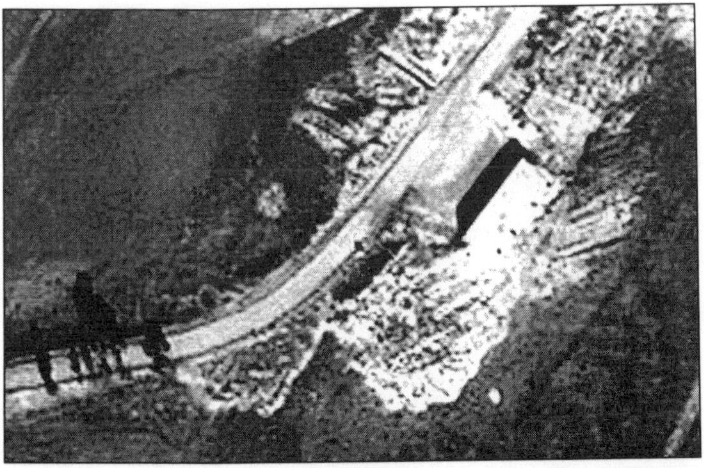

Figure 5.20

Missile bunkers are often dug into the side of hills. This Syrian Scud missile bunker near Hama, Syria is built in a narrow valley so Israeli aircraft will find it hard to bomb. (Space Imaging and CBN News)

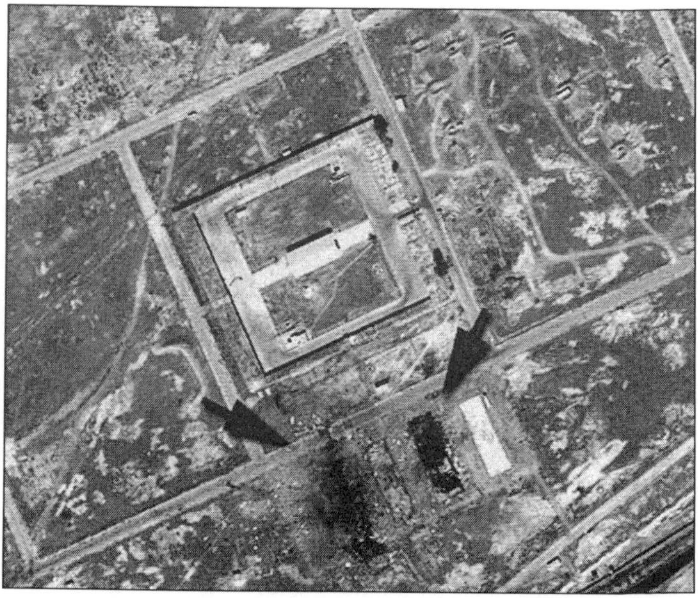

Figure 5.21

Scarring from bomb damage is easy to see. The arrows point to two buildings destroyed by U.S. air attacks in Basra, Iraq in 2003. Just above them, you see another drive through building indicative of a missile facility. (Space Imaging)

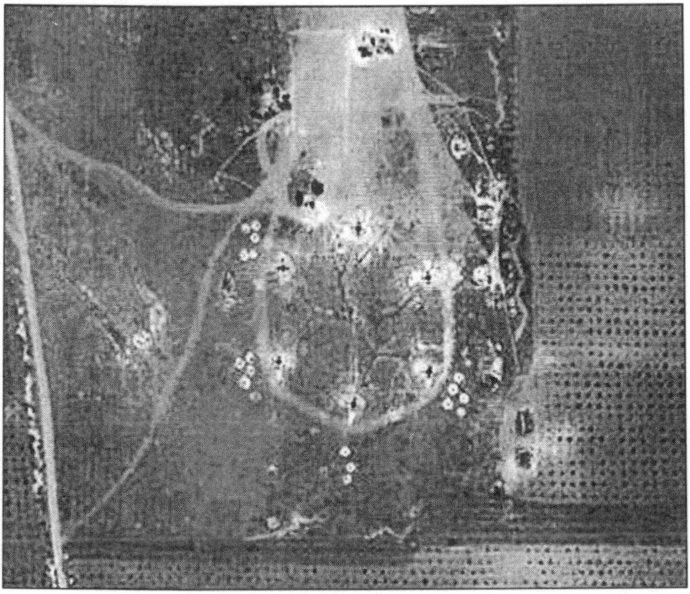

Figure 5.22

An anti-aircraft artillery (AAA) position. Like a surface-to-air missile site, it is circular, but the artillery pieces are smaller than missiles and look more like howitzers. (Space Imaging)

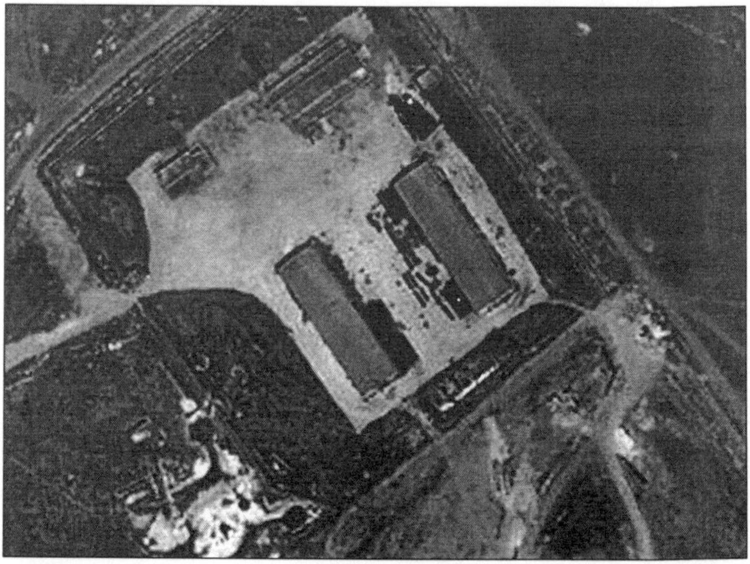

Figure 5.23

Although tanks, missiles, and other weapons can be hidden, the infrastructure that supports them is harder to hide. This warehouse at a Syrian Armored division is holding hundreds of drums. It is quite possibly a fuel and lubricant depot. (Space Imaging)

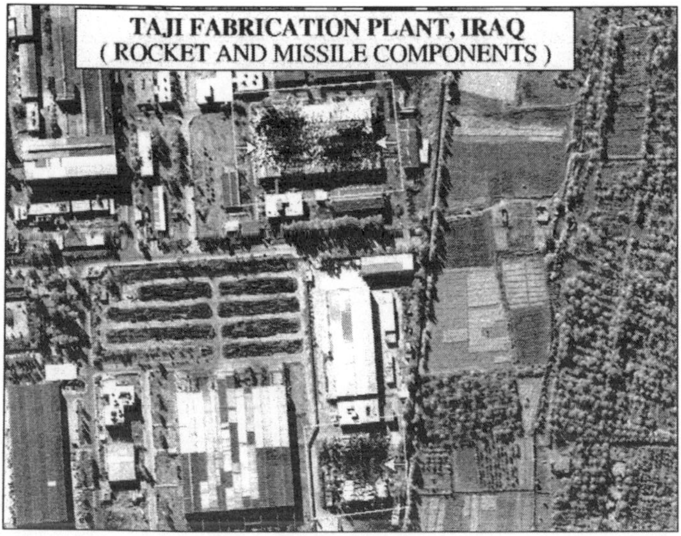

Figure 5.24

Many critical military installations are hidden from satellite interpreters. This image of the Taji Missile Component Fabrication Site looks like a normal industrial facility. In that case, additional intelligence is required. (DoD)

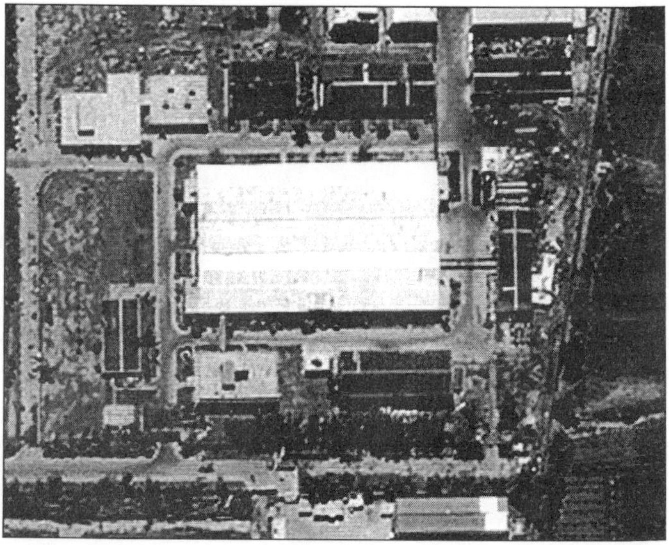

Figure 5.25

Comparing imagery over a period of time is critical to interpretation. In this image, you can see that the building damaged in figure 5.24 has now been repaired. (Space Imaging)

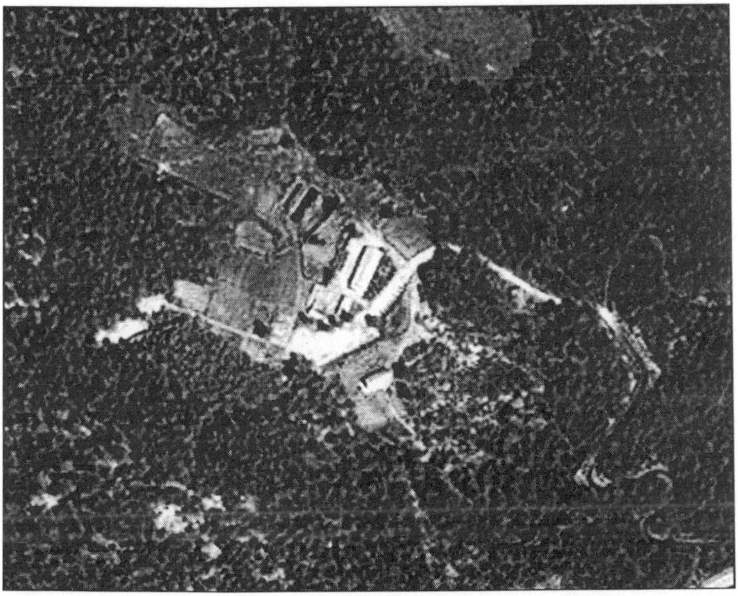

Figure 5.26

Ask questions of yourself. Although the North Koreans had promised to stop nuclear weapons development, this facility in Yong Byon, North Korea was built within walking distance of the closed plutonium separation facility. This facility was hidden in the woods so it wasn't visible to international inspectors. It is possibly a facility to extract plutonium from wastewater. (Space Imaging and CBN News)

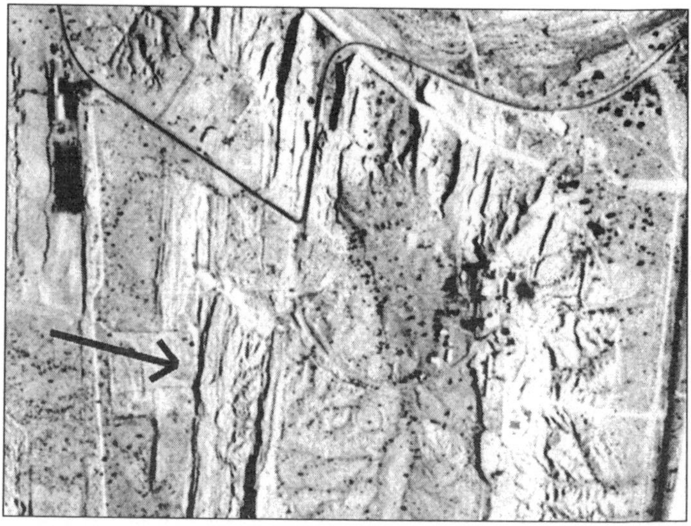

Figure 5.27

Always keep an eye open for underground facilities that have a security perimeter. This Russian KVR-1000 image of Bandar Abbas, Iran shows the Iranian Air Defense Command. The clues are a road leading underground and a security perimeter. (Sovinformsputnik)

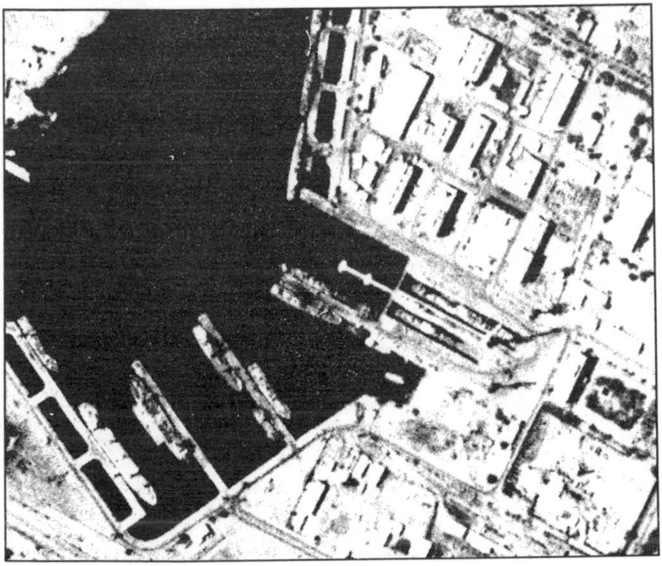

Figure 5.28

The Iranian naval facility at Bandar Abbas. While merchant ships have wider beams in order to hold more cargo, military ships are narrower and built for speed. Can you tell which are merchantmen and which are warships? Can you see the dry docks? (Sovinform-sputnik)

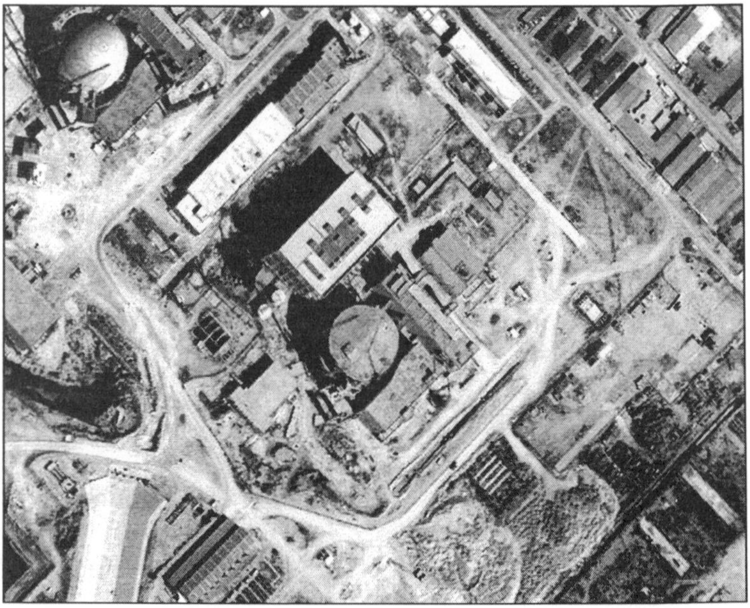

Figure 5.29

Two nuclear reactors at Bushehr, Iran. Nuclear reactors are easy to identify because of the large containment vessels (center and upper left hand corner). The reactor in the center is under construction. (Space Imaging and CBN News)

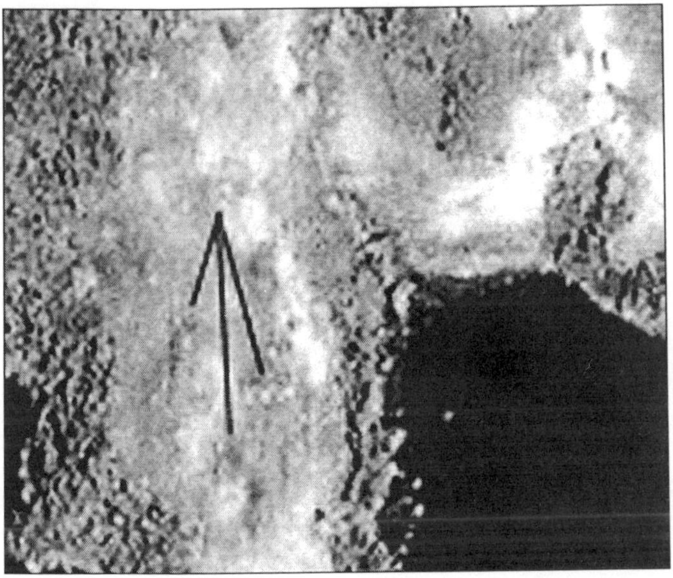

Figure 5.30

Camouflage makes interpretation difficult and often requires patience. Three camouflaged buildings related to North Korea's nuclear weapons program are barely visible in this image. Shadows and paths that lead to camouflaged structures are sometimes the best clues. (Space Imaging and CBN News)

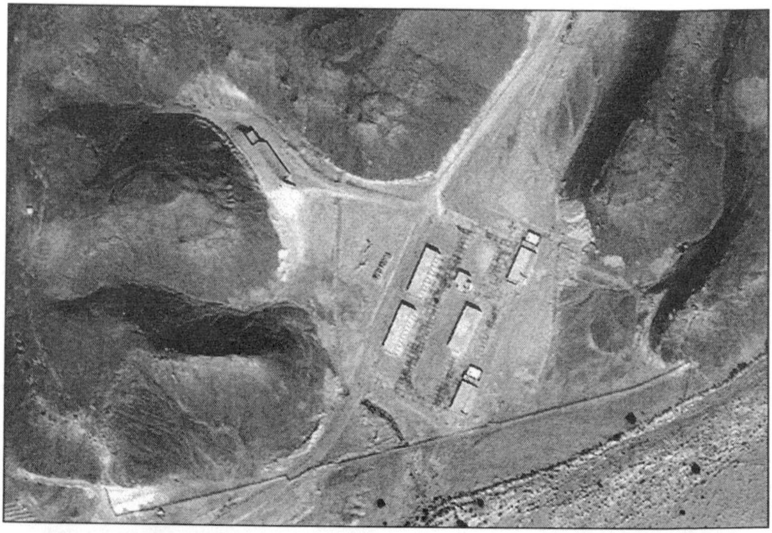

Figure 5.31

Commercial satellite imagery is excellent for military planning. This image of a ballistic missile maintenance facility can be used to determine infiltration routes, sniper positions, enemy strongpoints, and distances. Ask yourself how you would carry out a raid on this site. (Space Imaging and CBN News)

Industrial, Commercial and Residential Areas

Although city planners have used satellite imagery for over a decade, commercial applications have been lacking. There, however, are many applications like marketing studies, real estate, and commercial development.

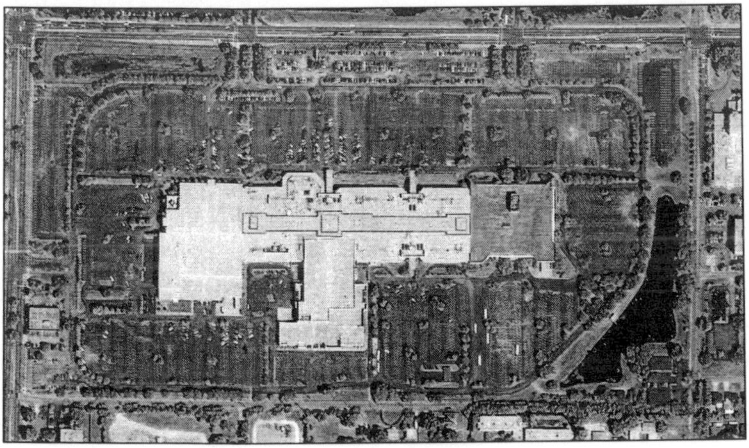

Figure 5.32

The typical American mall. A look at the parking patterns shows which stores are doing well and which are suffering. Since malls often precipitate more commercial development in the area, a potential real estate investor could use satellite imagery to analyze where future stores may go and what property might be a good investment. (Space Imaging)

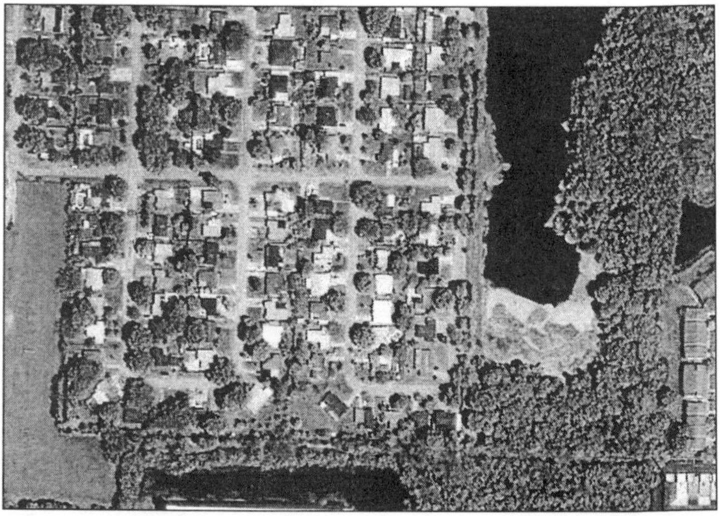

Figure 5.33

Satellite imagery of a residential area can give the viewer a good idea of the type of residents and their income. These views are also good for selling a property because they show amenities like this lake. (Space Imaging)

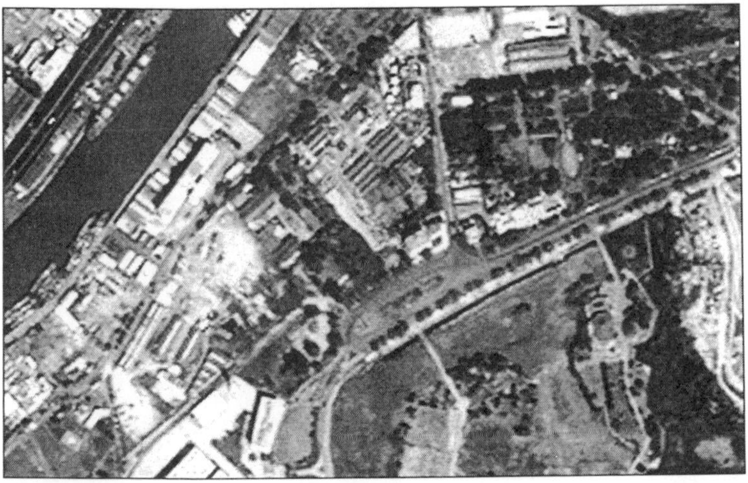

Figure 5.34

Commercial port in Buenos Aries has warehouses, as do most older ports. (Space Imaging)

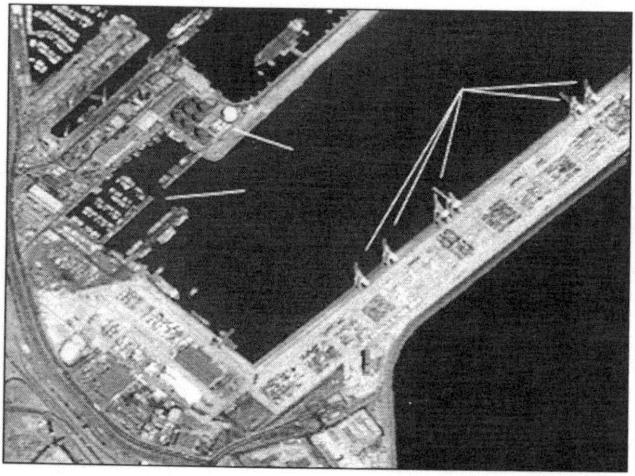

Figure 5.35

Modern ports like Cape Town's have fewer warehouses since the goal is to move goods quickly out of the port to the final destination. Here five cranes can move containerized cargo from the ships and load it onto railroad cars. Also note the marina and the oil tanks. (Space Imaging)

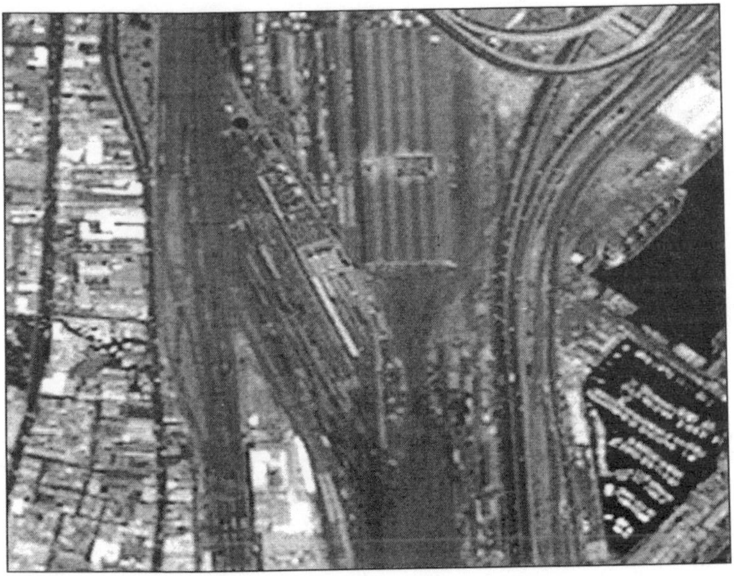

Figure 5.36

Rail terminal at Cape Town's port. Rail lines are obvious because they have much wider turns than roads. (Space Imaging)

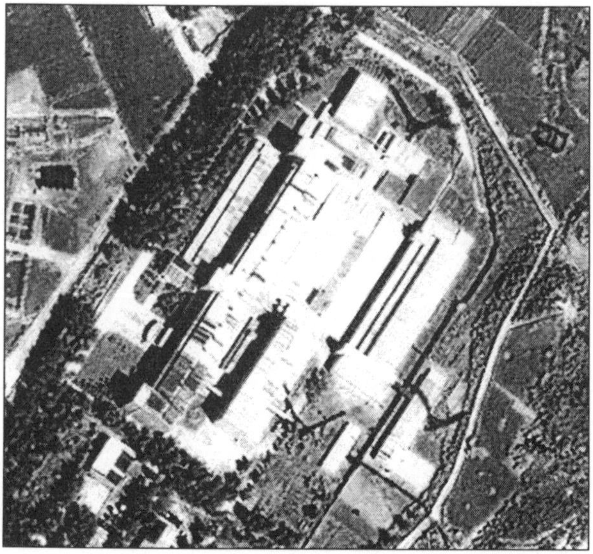

Figure 5.37

Each type of industrial process has distinct characteristics that can be seen by satellite imagery. Keys to look for are raw materials outside, arrangements of chimneys, stacks, boilers, tanks and cranes. If you are lucky, you can see the finished product in the yard waiting shipment. This is a light fabrication industry plant, as evidenced by a light steel frame, multi-story building, lack of heavy lifting equipment, and little open storage of raw materials. A heavy fabrication plant would have a heavy frame, be only one story tall, have storage yards for heavy lifting equipment, open storage for raw materials, and probably a rail line nearby. (Space Imaging and CBN News)

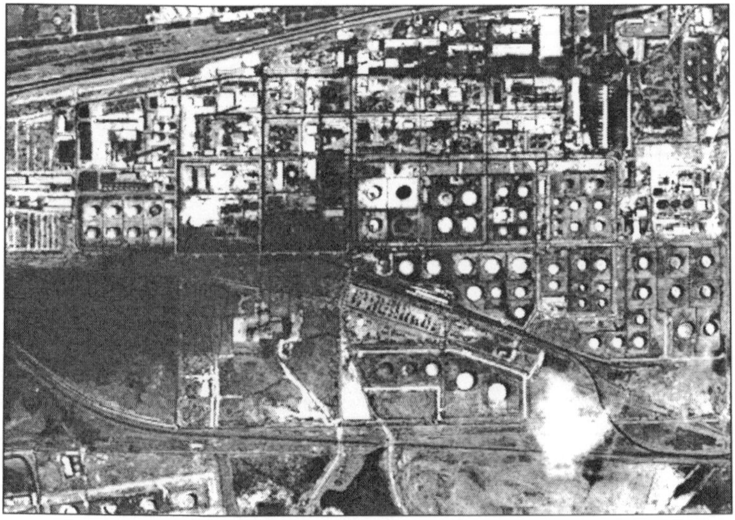

Fig 5.38

Processing industries are obvious because they have storage facilities for large quantities of bulk materials (ponds, piles, tanks, etc.). There are also usually outdoor processing facilities like kilns, blast furnaces, cooling towers, cracking towers, etc. Since a lot of energy is used, there are signs of coal, electrical power, chimneys, etc. In the case of this refinery (chemical processing) one sees tanks, pipelines, and outdoor processing equipment. Heat processing industries have large stacks or chimneys, kilns, and large quantities of fuel like coal. In mechanical processing, there are few outside signs like kilns, or stacks. However, there may be a transformer yard nearby. (Sovinformsputnik)

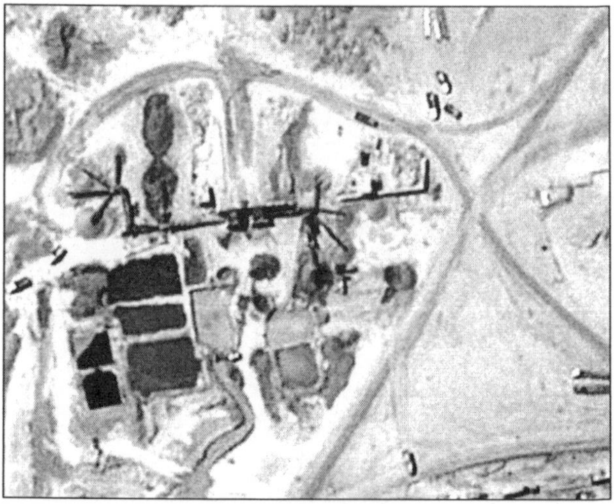

Figure 5.39

Extraction industries are characterized by excavations, ponds, derricks, waste piles, stored materials, conveyors, earth moving equipment, etc. This gravel plant that has been built on the site of Iran's nuclear facility at Natanz, Iran has many of those characteristics. (DigitalGlobe)

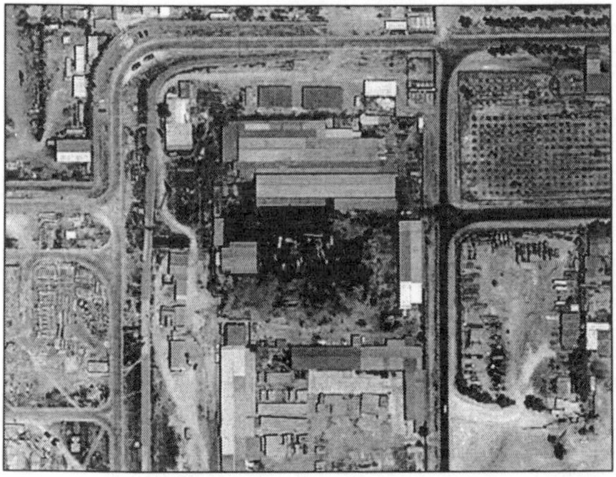

Figure 5.40

Vehicle maintenance areas like motor pools are often indicated by stained blacktop like this motor pool in Bushehr, Iran. (Digital-Globe)

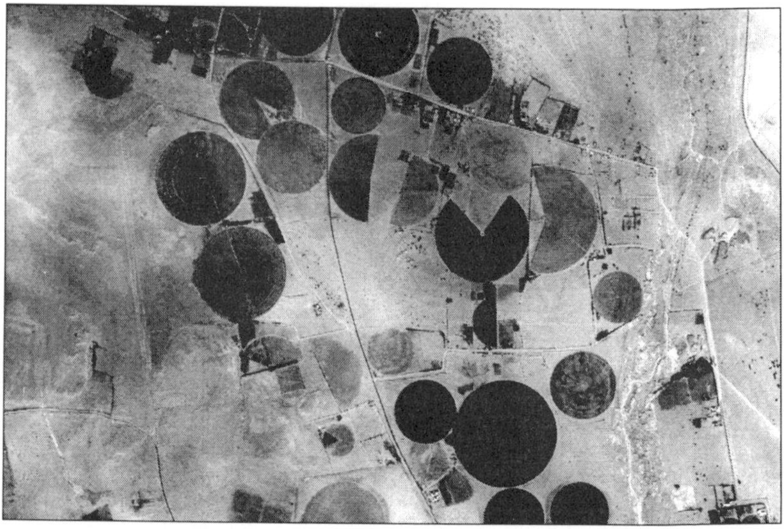

Figure 5.41

Irrigation circles in Saudi Arabia. Farmers who use satellite imagery can detect problems with their irrigation equipment like clogged sprinklers by looking for circles of stressed plants. (Space Imaging and CBN News)

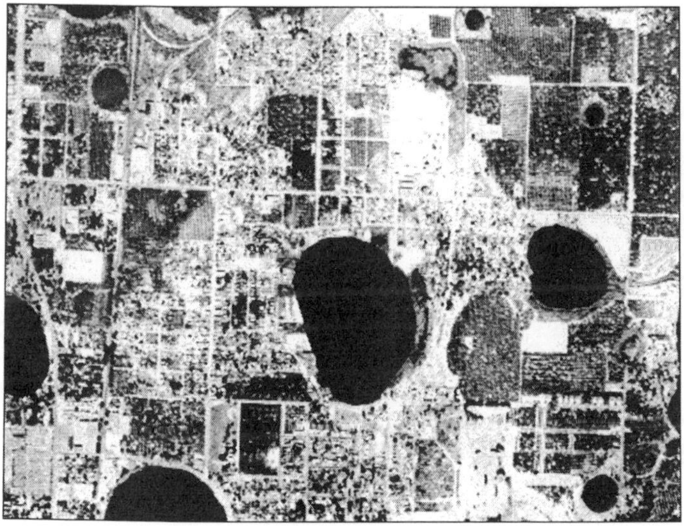

Figure 5.42

A limestone area, as indicated by the lime sinks (lakes). This is the type of region that is likely to have caves. (NASA)

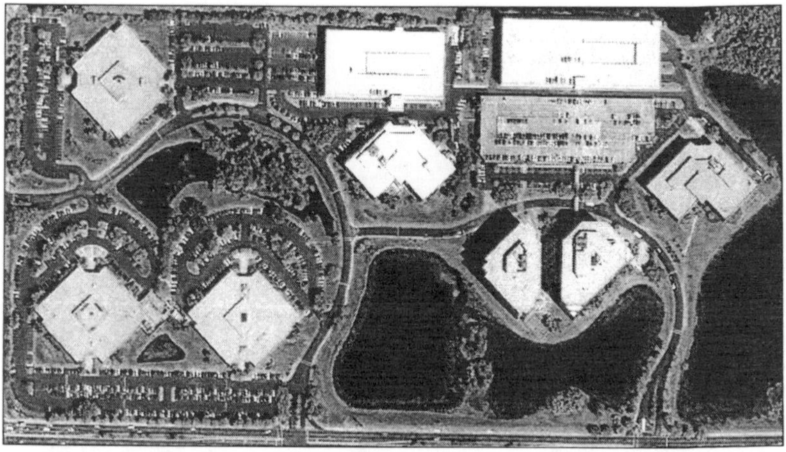

Figure 5.43
Modern office buildings have large parking lots and extensive landscaping. (Space Imaging)

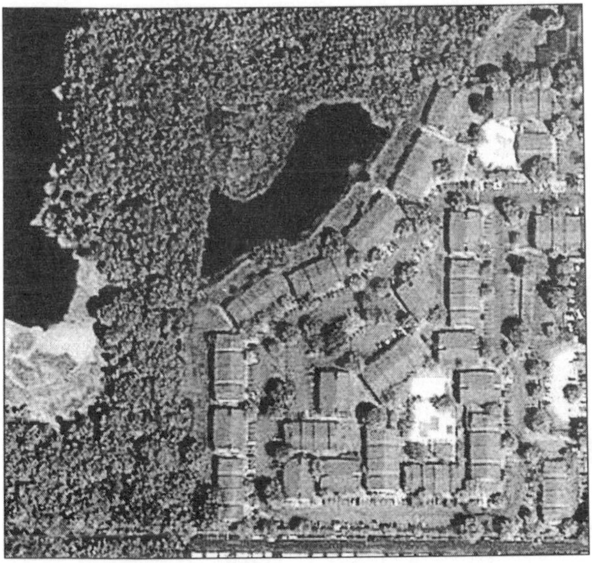

Figure 5.44

A typical American suburban apartment complex. Several buildings with parking space around each building. Attention is paid to landscaping. (Space Imaging)

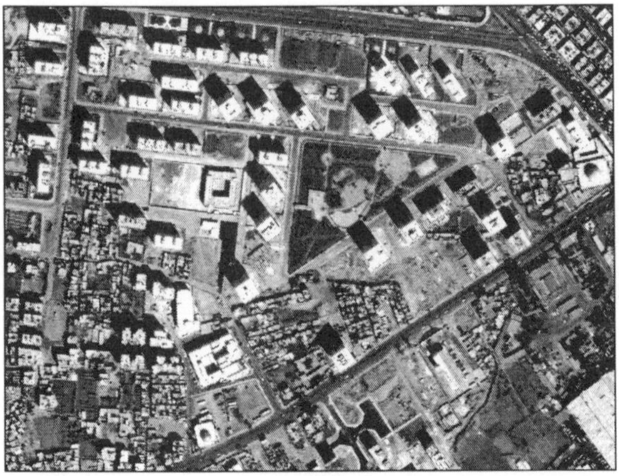

Figure 5.45

Apartments in other countries usually don't have as much parking space as in America. These Damascus apartment buildings have little parking and fewer outdoor amenities. There is just one park for over two dozen high-rise apartment buildings. (Space Imaging and CBN News)

Chapter Six
How to Make Money With Satellite Imagery

So far we have seen how governments, major corporations and experts use and profit from satellite imagery. But, how can the average man use this new technology? Well, here are some ideas that you can profit from.

Gold Prospecting

Geologists have used satellite imagery to find oil and minerals for years. Usually, this imagery analysis is beyond the average person, but not when prospecting for gold placer deposits, the type mined by the Forty-niners during the California Gold Rush and modern amateur gold prospectors today.

Alluvial placer deposits are those formed by the action of water on gold bearing ore. Gold, being one of the heaviest naturally found elements (seven times heavier than sand) will naturally settle out of the water whenever there is an area of slow moving water. That is where satellite imagery comes in.

With an image, you can quickly cover miles of stream and river courses to find the most likely places for placer deposits.

One of the best targets for placer gold are the bends of the river. Usually the fastest current is on the outside of the river bend, with the slowest current on the inside. Naturally, heavy material will drop out of the water at that point. In many cases, this is very evident because there are sandbars at these points. Again, the bright white of the sandbar is evident from space. Since gold is heavier than the sand, the potential gold deposits will be on the upstream side of any sandbar.

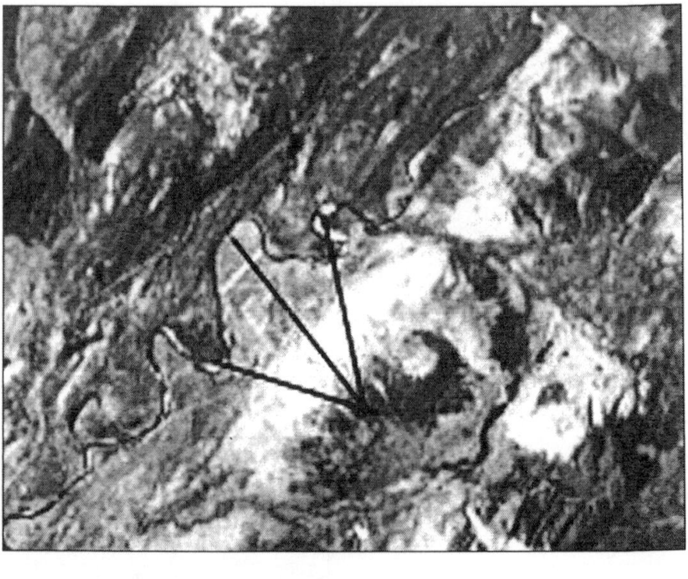

Figure 6.1

The curves of this river are natural places to look for placer gold. Note that white sand bars are also evident. (Sovinformsputnik)

Not all gold is to be found in the river. The greatest water movement is during flood stage and satellite imagery is excellent for discovering the flow of the river at flood stage, often because there is less vegetation in the flood bed. As with regular rivers and streams, the best place to look is the inside of the flood bend.

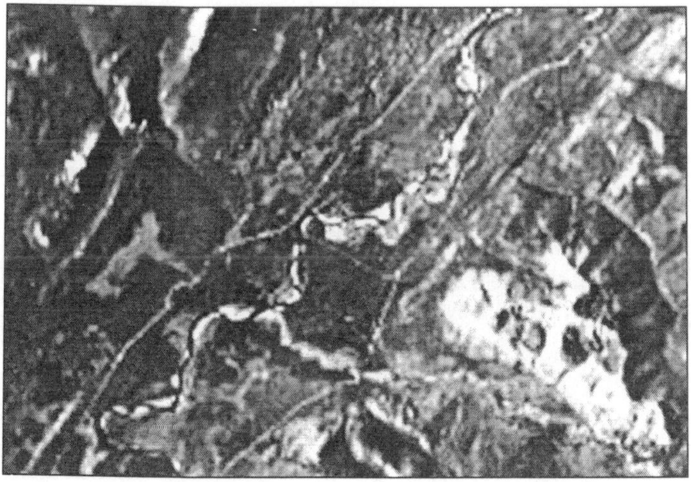

Figure 6.2
The river has changed course over time. Satellite imagery can be a guide to potential placer deposits further inland, away from the river. (Sovinformsputnik)

Bends of rivers and streams aren't the only places gold drops out of the water. Any place where water speed goes from fast to slow is a good place to check. For instance, when a narrow channel of water suddenly widens, the water slows down. The same is true when a shallow riverbed suddenly becomes deep. With visual band satellite imagery, you can not only see the water line, but also have a better feel where the riverbed becomes deeper.

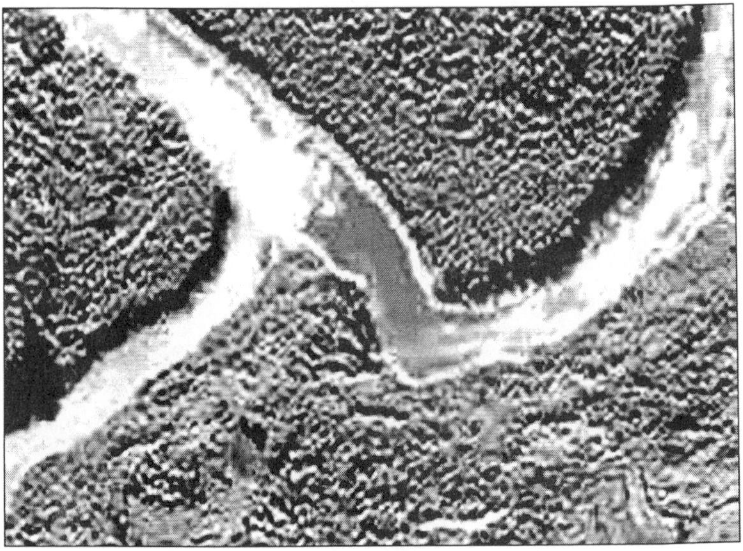

Figure 6.3

Gold placer deposits are likely where water speed changes like when a narrow tributary goes into a larger river, with a slower water speed and when the river goes from a shallow to deep river bed. (Space Imaging)

Similar placer deposits can also be found on beaches of oceans and lakes (Nome, Alaska's gold was found on the beach). The same principle can be used for finding other minerals like platinum nuggets or even diamonds.

Treasure Hunting

Treasure hunting is a big business, as evidenced by the growing popularity and variety of metal detectors. It seems like everyone is walking around with a metal detector, looking for lost coins, jewelry, and watches.

Most metal detector prospecting is hit and miss — scouting popular places like beaches and parks. However, one doesn't have to go to New Mexico or Saudi Arabia to use it.

The American West was the scene of one of the greatest migrations just 100 years ago. Hundreds of thousands of pioneers went west on horseback or by wagon over the ecologically fragile Great Plains, Rocky Mountains, and American Desert. In many cases, these trails are still visible to the naked eye. In the case of trails that have appeared to be grown over, they are still visible to either panchromatic or infrared imagery because much of the vegetation in the West doesn't heal as quickly as it does in the Eastern U.S.

It was not uncommon for pioneers, who were overloaded, to throw things off on the side of the road as they traveled. Often times, someone picked these objects up later, but many times, they were covered up by dirt and vegetation over the years.

Today, much of this antique material still rests on the side of old trails that have been lost over the century. The only things needed to find it are a metal detector and a piece of satellite imagery.

Rather than just randomly pick some satellite imagery, it would pay to do some research first. Since famous trails like the Oregon Trail are well marked and picked over, the best bet is to go with some minor trails. A good bet would be abandoned trails that go to ghost towns.

There are several popular books that describe ghost towns and how to get to them. When you find a likely one, acquire some satellite imagery of the area (preferably panchromatic or infrared). In arid areas, the trails should still be visible. If they are hard to find, look for natural "choke points" where the trails would have to pass (passes or saddles in mountain ranges, natural fords in rivers, easy grades, etc.). Then you can find the trail and follow it.

Although the whole trail is a likely treasure hunting area, there are choice spots to begin your hunt. These are at rivers, by steep grades of the trail, and any other place where pioneers might have dumped gear because the additional weight made moving the wagon difficult. Another potential spot would be former campsites, which often were flat, near water, and shade.

Although you probably won't find any really valuable treasures, this kind of treasure hunting should give you a steady stream of small antique metal items that would be of interest to antique collectors.

One final note. State, county, and federal law often regulates hunting for antiquities. Be sure you know the rules before you start your treasure expedition.

Treasure Hunting Part II

If horseshoes, nails and bits of metal early Americana aren't enough for you, you can use satellite imagery to look for bigger treasure. As we mentioned in earlier chapters, satellite imagery has been used in the search for sunken Spanish treasure.

There is a lot of "lost" treasure in the world. Whether it came from miners, pirates, or whoever, there are still treasure troves around for those smart enough to find them

Obviously, someone can hide treasure anywhere, however, they are more likely to find a hole and cover it up. That leads us to limestone formations. Satellite imagery can identify limestone formations, where caves are more likely to be found. Although this imagery can't pinpoint a specific cave, much less tell you if it has anything in it, it can give the prospective treasure hunter likely areas to cover.

If you are looking for buried pirate treasure, you can limit your search area through the same process of elimination. A pirate treasure would be more likely to be buried near a beach than cliffs, where the boats couldn't land to unload the treasure. There would also probably be a good anchorage for the ship (deep water as evidenced by blue light), and a good lookout for watchers.

Whatever the treasure you are looking for, satellite imagery can help you if you use your brain and take the time to think it through.

Deer Hunting

Okay, this one won't be a moneymaker (unless you are a professional hunter or guide), but it can help you bag your deer.

Like many animals, deer tend to graze in open areas, where tender grasses are found. If you intend to hunt in a new area, a satellite image can help you find open areas with young, tender grass and watering holes. Of course, satellite imagery can be used to find other game and it may even help you find your way back to the car if you get lost.

Mapping

Satellite imagery is a popular tool for governments and private groups involved in mapping. However, you too can use satellite imagery for mapping.

If you live in a small town that doesn't warrant a map, if you want to map a favorite wilderness area, or whatever, you can produce an overhead view/map with digital imagery and a computer.

Investigative Journalism

So far, most satellite-based investigative journalism has been defense- and intelligence-oriented. Yet, there is still more that can be covered. True, the "Axis of Evil," Iran, Iraq, and North Korea have been well covered and reported on with satellite imagery, but there are many other countries that have been barely touched; Syria, Israel, India, China, and Cuba. Then, there are all the ongoing conflicts in the world that can be covered: Chechnya, Rwanda, the Congo, Sudan, etc. Obviously, there is more than enough to keep everyone busy covering the world's death and destruction.

But, investigative journalism isn't limited to that. One area that has barely been touched is environmental reporting — on both sides of the issue. Did that company outside town actually dump that toxic waste (remember that near infrared imagery will show sick plants)? Is the Amazon rainforest really disappearing at the rate claimed (there is evidence that the environmentalists are over estimating damage)? Has the habitat of the Lesser Crested Red Tipped Australian Wopperwill really been devastated by development? Satellite imagery can provide conclusive evidence.

Other areas where satellite imagery can be useful in investigative journalism are city planning and development, zoning issues, new road construction, local environmental issues, business development, etc. The only limit is your imagination.

Computer Games

What was it like to bomb Hanoi in 1967? Or Moscow? With declassified satellite imagery from America's spy satellites you can add a digital image of Moscow, Hanoi or whatever city to an air combat computer game. As you will discover in a later chapter, making three-dimensional "fly-throughs" is easier than you think, thanks to modern computers.

Nor do you have to limit yourself to games from the Cold War. Maybe you want to see what a Civil War battleground looked like before it was overrun by development. By using either spy satellite or old aerial imagery from the National Archives, you can turn the clock back to a time when the battlefield was still pristine. The same holds true for World War Two and archived aerial intelligence imagery of the era.

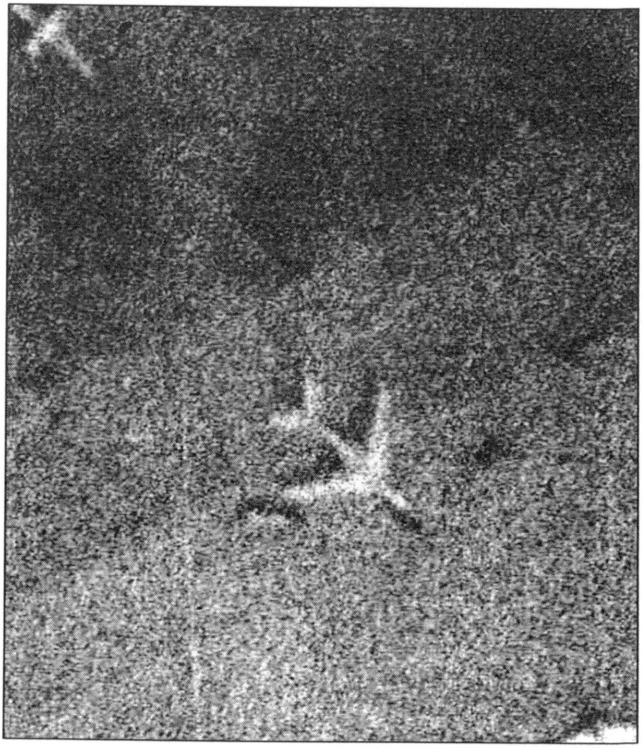

Figure 6.4

An aerial photograph with an interesting history, from the first U-2 flight over Moscow on July 5, 1956. The primary mission for this overflight was to discover how many jet bombers the Russians had. This photo shows one of the few jet bombers found at Soviet airfields.

Consulting

Satellite imagery offers any number of possibilities for consultants. How many swimming pools are in a neighborhood (what percentage are kidney shaped)? How many yards are fenced (yes, you can tell that with a 2 foot resolution satellite image)? How many have barbecue grills outside? All of these can be done quickly with a satellite image of a neighborhood.

If marketing is your bent, better yet. How many houses are within a certain radius of a potential hardware store site? How many people with gardens are near a plant store and what neighborhoods are they in? You can even make a good guess as to how many acres are under cultivation in a suburb. Do people with garages still park their cars outside (which means the garages are filled with consumer goodies)? You get the idea.

Consulting isn't just limited to residential areas, although they are a good target for marketers with money. Any business that intends to open a store needs detailed marketing information and a satellite image database provides more info than the conventional marketing study because you can go back to learn more. For instance, a marketing survey of a farming area would use information from the Department of Agriculture on acres under cultivation, crops, etc. But, it wouldn't show if farmers are showing signs that they might be optimistic about the future and expanding their operations. A marketing firm could go back to the satellite data and look for signs of expansion like construction of new barns or tilling fallow fields. In that regard, a satellite image is a database that keeps on supplying information long after the traditional marketing survey is finished. In fact, since a single farm can be viewed, a farm equipment retailer could check out the farm of a potential client to see what the farm and land looks like, giving him an idea of what the farmer might need.

Real Estate

Before packing your clients up in the car for a long tour of houses, spend a little time in the office looking at the prospective neighborhoods. You can give them a bird's eye view of their potential home, show them exactly where the nearest grocery store, school, playground, and interstate turnoff is before they visit. That in turn may tell them that it isn't the right place for them (which saves you time) or helps close the sale (which helps you make money). Since satellite imagery costs hundreds of dollars and the cost can be shared by several real estate agents, it is a real money saver in the long run.

Satellite imagery is also a potential tool for real estate developers. Although surveys and engineering plans will be required, a developer can buy a satellite image of a potential property and make plans long before giving it to the engineers. The result is that the developer can determine if the property is one he wants to develop and a lower bill from the engineers because he gave them better information (based on satellite data) in the beginning. The imagery may also be critical for making presentations to the city and winning their approval.

Coming up with satellite imagery prices is like coming up with computer prices. It all depends on when the image was taken, resolution, image size, color, number of bands, the format, etc. You can get free images on the Internet or spend about $10,000 to $20,000 on a scene if you place a special order that may require a change in orbit or priority work. However, prices would probably be in the neighborhood of $200 to $1,000 for good, high resolution data, depending on how large a scene is. But, you can do a lot with free imagery, if you know what you are doing.

Fighting City Hall

Let's say the current "anti-gun" government has announced that they will be closing the local shooting range because it poses a risk to a planned building development, there is an environmental threat from lead, and local residents complain about the noise. Rather than take it lying down, you can fight back.

Satellite imagery not only makes a persuasive tool for a presentation, it also includes information that can be used to refute the politicians. You can measure the distance to the new development that is supposed to be threatened by the range. Ballistic tables can tell you if it is even in range of a bullet. Infrared imagery can show if the vegetation shows any more signs of stress than the mayor's yard. A little research can give you the calculations for noise attenuation so you can calculate the actual level of noise at the local houses. You can even ask if trees can be grown to dampen noise rather than closing the range.

A piece of satellite imagery, a good presentation and a good talker (not a yeller) can probably stop the politicians or at least stall them long enough for you to get more support. If nothing else, the presentation will probably garner a positive newspaper story and maybe give you some contacts for moneymaking ventures.

Writing

A good writer usually visits the setting of his book. That's a good idea, unless you are writing a spy thriller set in a top-secret military facility. Of course, you can fake it, but the results are never as satisfying, and one of your readers may have visited the area.

If you can't visit the locality, why not buy a satellite image of the area? For less than the cost of a plane ticket, you can visit any place in the world and gain a familiarity with the place that will make your book seem more realistic.

If you want to gain more perspective, you can actually create a three-dimensional image that will allow you to know what can be seen from the top of a hill or what you can see when leaving a certain building. You can even take a virtual walk through the city without leaving your house.

Investing

We have already discussed using satellite imagery to monitor crops, which can be useful in commodities investing. However, satellite imagery can be used in more than that. Satellite imagery has helped investors acquire information about operations in closed societies in the past. For instance, platinum production in Russia is still considered a secret. With satellite imagery, an investor can track production at the mines by counting ore cars, how much earth is moved, and smoke stack emissions.

Security

In this era of increased concern about terrorism, plant security has become a high visibility issue. Now, with satellite imagery, the security department or an outside group can carry out a security perimeter audit.

A high resolution image can display the perimeter of a secure facility and show where potential loopholes may exist. It will show buildings and offices that may be vulnerable, trails outside the perimeter that may come too close to the fence, and high points that observers or snipers can use. In the case of high security areas like nuclear reactors, you can map out the whole area that may be inside rifle range and make changes to your security that will prevent any terrorist attacks. An up-to-date satellite image would also be ideal for security patrols and the security office.

An excellent way to get a security contract with one of these potential targets is to buy some imagery, think like a terrorist, and figure out how an outsider can cause trouble. Then take some of these ideas, produce a professional presentation, and approach their security department with a proposal that you produce a report on their security weaknesses. If you do your homework and come across as a professional, you can probably show them that they need to hire you as a consultant.

Chapter Seven
Acquiring Satellite Imagery

So far we have talked about what satellite imagery is, what it can see, and how to make money using it. I guess it's about time to talk about how to acquire it.

It's a lot easier now to get satellite imagery than it was even ten years ago. First, there is a highly competitive market in satellite imagery, which forces the various satellite owners to keep their prices low. For instance, just ten years ago, LANDSAT imagery (with a resolution of 30 meters) cost over $4,000 a scene! Today, you can get the same or a newer scene from LANDSAT for free or a nominal fee (about $10) from U.S. government agencies.

The second change in satellite imagery purchasing is the Internet and the option of buying it online immediately. What was once expensive and often took months to get, is now cheap enough to put on your credit card and download immediately. There is also a considerable amount of free imagery that can be just as useful and even more interesting historically.

In this chapter we will briefly go over the various options. In the case of commercial companies and some government sources, I will remain vague on details because this is a dynamic industry that changes regularly. Much of their business is done online and their web sites change frequently. In addition, satellites are always going up and breaking down. Any list of available satellite systems would be outdated before this book goes to printing. However, the information in this chapter and the contact information in the Appendix will let you get anything you need.

Before You Start

Before investing the time necessary to find the imagery you need, you should do some homework.

Location. It may seem obvious, but you need to know exactly where the scene is. In the case of lower resolution scenes that often cover many hundreds of square miles, the exact location isn't as necessary, but if you intend to use higher resolution imagery and don't want to pay for a large image, you need a location that is accurate.

The best way to fix the exact location is with latitude and longitude. But you need more than the latitude and longitude in degrees. Latitude and longitude are graduated in degrees, minutes, and seconds. One degree is broken into 60 minutes and one minute is broken into 60 seconds. If you want accuracy to within one nautical mile (6,076 feet or 1.151 miles) of the target, you need the exact latitude and longitude within one minute of a degree. If you need an accuracy of 100 feet, you must get within one second of a degree.

If you are able to physically visit the site, the best tool is a Global Positioning Satellite (GPS) receiver. These can give you the necessary accuracy for about one hundred dollars. If you can't visit the site, it is a bit harder. However, the U.S. Census Bureau has a web site that allows you to obtain a latitude and longitude by just giving a zip code. Of course, if you live in a sparsely populated area, the zip code position will not be very accurate.

Maps, like those from the U.S. Geological Survey (USGS) are very accurate and give you the latitude and longitude. If the target isn't inside the U.S., nautical or aeronautical maps can be used. These are often found at university libraries. If that isn't an option, most libraries have what is called a gazetteer, which gives latitudes and longitudes of most cities and towns in the world. If you can't get an exact position for your area, you can at least get the location of a nearby town or city that will help in finding the exact location.

If you are looking for a specific point like a house, a simple latitude and longitude is enough. However, if you are looking at an area, you will need to identify what are called "corner points." These are the four corners of the scene that you wish to acquire. The area of interest will fall inside the four corner points.

Once you have a position, you can use the extensive resources of the Internet to define your search. The exact places you should go are in the Appendix, but basically, here is how you can use them.

Visit one of the satellite imagery web sites that offer low resolution views of their imagery. With your latitude and longitude, you can get a list of the imagery they have of the area and a low resolution view that will give you an idea of quality and cloud cover. It should also give you enough information to know if your location information is correct. If what you are looking for isn't in that scene, be sure to check the adjoining satellite images.

Before you begin your satellite search, do some research on what you are looking for. If, for instance, you are looking for a secret missile site in North Korea, make sure you know what you are looking for. Be aware of critical signs of missile bases like large radius roads, bunkers, launch sites, etc. Then research the web to see if others have used satellite imagery to locate the facility (see the Appendix for several web sites that specialize in this field). Take time to read all you can about the site, so when you begin your search, you will spend less time and make fewer mistakes.

These low resolution views won't tell you everything. If you are looking for treasure, the image should tell you if the area of interest is in the scene, if it is cloud free, and if potential signs like vegetation might be good for your search. But you will still need to buy the image to get the additional detail.

These low resolution scenes are also a good way to lower your costs. Many satellite imagery companies don't require you to buy a full scene, just a piece of it. With a low resolution scene you can tell them exactly what you need. Rather than tell them that you require a certain scene, you can give them a scene number and an image of the scene with your requirements marked on it. Some companies will then give you the smaller piece of imagery and just charge you by the square kilometer. Your costs will be lower.

The National Archives

If your interest is history, there is no better place to go than the National Archives near Washington, DC. Much of this imagery will also be available at the EROS Data Center, but the

National Archives is still the shoebox of the nation, in that all those pieces of paper are kept there. Not only will you find spy satellite imagery, but U-2 photos from the Cold War, aerial imagery from World War Two, and other overhead photographs of interest to the historian. Since the land was less likely to be disturbed by mankind several decades ago, it is also great when searching for archaeological sites.

Although the National Archives in downtown Washington is best known because it houses priceless pieces of our history like the Constitution and Declaration of Independence, most of the material is kept at the National Archives in College Park, Maryland, a suburb of Washington. This modern facility opened in 1994 and can be easily reached by public bus, Metrorail, and free buses from many of the nearby hotels, who cater to researchers (the Archive has a list of local hotels that have shuttles running to and from the facility). The Archive is open from Monday to Saturday, but Saturday is a poor day for research on satellite imagery, because they will not go into their files to get imagery. It has to be pulled on Friday, so you can view it on Saturday.

Although the records are open to the public, you will need to register as a researcher, which will take a couple of minutes in the beginning. Needless to say, they do restrict the type of materials taken into and out of the archive areas. Notebooks aren't allowed, although paper is available inside the area. Cameras are allowed and probably necessary for any satellite imagery study. Since much of this material is one of a kind, they are careful to make sure that it doesn't "disappear."

Although you can order copies of satellite imagery through the National Archives for a nominal fee, it is best to bring your own camera, so you can immediately copy it. A good 35mm SLR is recommended. You will also need a GOOD macro lens. The type found at most camera stores is more than adequate for snapshots, but not for accurate copying of satellite imagery. A heavy, fixed focal length lens especially made for macro photography is a must if you want to get the most out of your trip. A pocket tripod and a device to remotely take the picture are also recommended.

Your film shouldn't be the same type you buy at the grocery store to take family snapshots. A slow speed, high resolution, black and white film designed for copying is best. If you need to compromise, go with a slow (100 ASA or better) color film. This may be cheaper, more available and can be developed at a local one-hour film developer. This can be critical if you plan to spend several days at the Archive. You can develop the film each night and go back the next day if there were any problems. For more information on what it takes to make good photographic copies, see my book, *Practical Guide to Photographic Intelligence*. It will give you more info on picking the right camera, film and lens for a project like this. It will also tell you how to take a good quality picture.

I do not recommend using a digital camera. A lot of the detail on the photographic film is so fine that any commercial digital camera can't detect it. If you have to, get a 35mm SLR camera before going because you don't want to waste the trip and effort.

Once you fill out the paperwork and gain access to the archives, you have to find the right imagery. Although the archivists are very helpful, here are the necessary steps to finding what you are looking for. The first step is to find the World Aviation Chart (WAC) of the area of interest. When you give the archivist the form with this number, they will return with a cartload of WACs that have a photographic mosaic of satellite imagery pasted onto it. This photographic mosaic, which was put together by CIA photo analysts, represents one mission. These photos show what the camera actually shot, the frames, the cloud cover, and overall quality.

With this information, you then ask the archivist for a book that converts the information from this WAC to a specific satellite mission. Then you once again fill out a form that will ask for the duplicate film roll of that mission. Since the KH-4 (J-1 and J-2) camera systems were the best, it is best to use these missions, if possible. Also pick the forward camera (there were two cameras for stereoscopic analysis) if possible because lighting usually made it the better image.

Once you get the film, you will go to a room with several old light table viewers that were once the property of the CIA. Again, you can get help putting the roll on the spool and viewing it. Then it is up to you to find what you are looking for.

When you have found the specific frame of interest, you are going to assemble the camera and tripod on the light table. The images are recorded on strips about 3 inches wide and 18 inches long. Each frame covers an area about 20 miles by 100 miles, depending on the satellite's altitude. The strips run perpendicular to the satellite's orbit and usually run from the NW to the SE or SW to NE depending on the exact orbit of the satellite. I can't stress the importance of having the right equipment, a sharp focus, and practicing a lot before going. Don't waste all of your work by messing it up here!

The most efficient way to work at the archives is to request all the WACs of all the areas of interest at one time first, then get all the film rolls at one time. Since each request takes time, it is much more efficient to request a lot of information at one time. It also means that all your photography will take place at one time, so you won't need to keep assembling and dissembling the equipment. An effective researcher can cover dozens of areas in a one or two day trip.

Since you will probably be spending a short time at the Archives, don't waste it by spending the time interpreting the imagery. Get the information and take the photographs you need while at the Archives. Then do your research when you get back to the hotel or home.

Here's a recommended 48-hour schedule for a National Archives researcher:

Day One. Arrive at Reagan National Airport in Washington, DC, in the evening. Take a shuttle to a College Park hotel.

Day Two. Take hotel shuttle to National Archives. Register and go to research area. Get the WACs of the areas of interest and get mission data. Fill out film requests before lunch. Go to lunch (they have a cafeteria on site). After lunch get the film rolls, review the necessary frames, and take photographs. Take the shuttle back to the hotel and take the film to a one-hour film developer. Review the developed photographs to see what needs to be retaken. Do a quick interpretation to see what additional information you need the next day.

Day Three. Ask for any film rolls you need to rephotograph or do more research. Leave at noon and head back to the airport for late afternoon flight.

Be sure to take copious notes of any research you do. If you need to revisit the National Archives and look at any specific frame, you can do it very quickly rather than going back over the same territory again. You can also use this information to get copies from the EROS Data Center.

Low Cost Imagery from the U.S. Government

The U.S. government's EROS Data Center (EDC) has been called, "the place to go to first," by satellite imagery experts. Not only does it have a worldwide archive, the imagery is either free or low cost, partly because it was acquired through government funding. A word of warning — don't look a gift horse in the mouth. The imagery isn't the best resolution, it isn't always new, and the data formats may not be able to be read by your computer. However, this is a good place to go first.

EDC calls itself, "a leading source of land information for exploring our changing planet." Part of their web site includes "EarthExplorer" which is a U.S. Geological Survey online data discovery tool that provides access to millions of satellite images, aerial photographs, digital data, and paper maps. All you need is the town name (or a zip code in the United States). The cost of getting this data is very inexpensive.

Since a NASA grant requires that satellite imagery be offered for free for a while, the databases on this web site can go from free to a nominal fee at any time. Resolution and quality depend on the system.

This site is also a source for U.S. spy satellite imagery for a low cost. However be warned. The cheapest prints are very low resolution contact prints and worth very little from a practical point of view. It is not a substitute for a trip to the National Archives. If you decide to use this source, spend the money for the best photographic products.

Another source for satellite imagery is the NASA web site. Again, although it is free, you are restricted in your choice.

What is available from these web sites changes on a regular basis. Rather than tell you (and rapidly become outdated) I recommend you visit their web sites frequently and look for yourself. Further information can be found in the Appendix.

Russian Imagery

Ten years ago, the Russians were leading the world in commercial high resolution satellite imagery. Today, they are "also rans." Although the Cold War is over and Communism is dead, the Russians don't understood capitalism or the declining prices of satellite imagery, and it shows in how they market their satellite imagery. Although their systems are competitive with Western commercial systems, they aren't as customer-oriented as Western companies. Needless to say, they were quickly blown away by the multitude of civilian firms in the West that offered the same services for a lot less.

Ironically, the Russians and an American partner were the first to offer satellite imagery online at a reasonable cost in the late 1990s. Because they respond so slowly (even with e-mail) regular users of satellite imagery rarely use them, unless they require some imagery of an area that Russia covers on a regular basis (usually former parts of the USSR).

However, if you want photographic products (rather than digital images), Cold War historical data, or a good price for older, medium resolution imagery, the Russians may have just what you want.

Contacts for Russian satellite imagery are available in the Appendix.

Commercial Satellite Imagery

The future of satellite imagery is online and the commercial firms are taking full advantage of it.

Unlike the U.S. government, commercial firms offer a wide variety of imagery online so you can quickly assess it, order it, and receive it.

Although each commercial firm is different, here is basically how you order the imagery. You go to their web site, where they have a search engine that gives you access to low resolution copies of their imagery. You can pick the scene either by giving the engine the name of a town or city or latitude and longitude. You can even pick the area by using your computer mouse to point to the area of interest on a map.

After an area has been picked, you can access a list of all the scenes that cover your area of interest. The information will include the precise location of the center of the picture, the corners, the time the picture was taken, cloud coverage, and precise information for ordering it.

Most commercial satellite imagery firms also offer low resolution copies of the actual image so you can see if clouds cover the area or image quality. If you are looking for a certain facility, this gives you a chance to see if your information is correct.

Depending on what you want, you can order online with a credit card and get nearly instant delivery via an e-mail or a File Transfer Protocol (FTP). In cases where the file is too large, you will have the image sent to you via a CD. Depending on the resolution, size of the scene, and type of imagery, costs are well below $100 a square mile.

Information on the firms that offer commercial imagery can be found in the Appendix.

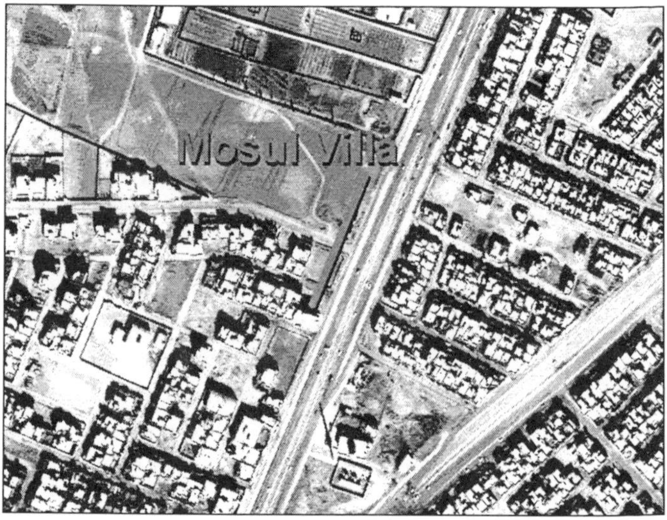

Figure 7.1

An example of the speed with which commercial firms move. This image showing the villa where Saddam Hussein's sons were killed by US Troops on July 23, 2003 was available the same day. (Space Imaging)

Chapter Eight
Satellite Imagery and Your Privacy

Let's cut to the chase. When it comes to satellite imagery, you have no privacy. With that in mind, let's look at who may use satellite imagery to invade your privacy, what they can do, what they can't do, how they do it, and most important, how to evade some of that snooping.

There are four major categories of people who will use satellite imagery to invade your privacy. They are:

Commercial entities. These are the people who want to sell something to you. For instance, they could use imagery to see if you have a swimming pool so they can try to sell pool supplies to you. Much of the information they gather may not even relate to you personally. They are more interested in characteristics of a neighborhood and demographics so they can decide if a store should be placed nearby. In some cases, a business may want to obtain imagery of a competitor in order to obtain business intelligence.

Individual. This can range from the person who wants to look at a neighborhood they are thinking of moving into, to the thief who wants to case a neighborhood or estate prior to a break-in.

Government. This is the scary one. Governments use satellite imagery for defense, intelligence, tax, nanny state regulation, and law enforcement purposes. The imagery can be used in court as evidence, and since the government has more interpreters and enhancement experts than anyone else, they can manipulate it in such a way as to serve their purpose. This is a growing concern, given the new powers granted to the government in the name of "fighting terrorism."

Government employees. These are government employees who illegally use government satellite imagery resources for their own purposes.

What Satellite Imagery Can't Do

Thanks to Hollywood, Americans have an inflated sense of satellite capabilities. The fact is that Hollywood writers and producers have no better idea of satellite capabilities than firearm capabilities. Just to set your mind at ease a little bit, here are some of the things satellites can't do:

Follow people running through crowds or city streets. Although the movie *Enemy of the State* was a good flick, with an interesting plot, spy satellites can't follow a running person through a crowd. First of all, the current resolution of spy satellites is about two to three inches. That doesn't give enough detail to differentiate one face from another, so the viewers could easily lose the person.

The second problem is tracking a moving person. Have you ever tried to videotape a children's soccer game? Remember how hard it was to track the player or the soccer ball flying in the air? Imagine trying to do the same from 500 miles in space. It's not that easy.

Then there is the problem of the satellite moving through space at thousands of miles per hour. As the satellite moves, the camera has to pan in order to stay on target. What was an overhead image becomes an oblique shot in a few seconds. That angle hides people behind buildings and even in crowds.

Satellites can see you in the dark and through clouds. Yes, satellites can do that, but the results are not the same. Radar can see through clouds and in the dark, but they are best used to detect good radar reflections like vehicles or some geological structures. In other words, they are great for tracking Iraqi tank columns, but worthless for following some person running through the rain forest.

Thermal imaging can also see in the dark, but it doesn't have the resolution of regular visual systems. They are much better at detecting the running engines of a column of tanks than an individual. They also work better when the temperature difference between the background and the object are great. In other words, a running tank engine in the winter is more noticeable than a person in the desert in the Southwest during the summer.

Thermal imaging systems are also nearly blind in clouds. If you remember from the earlier chapter, infrared energy is absorbed by water. And since clouds are made of water...

Enhancement Techniques can sharpen an image so you can recognize a face. There are a lot of enhancement techniques, but none that can sharpen a satellite image so you can recognize a face (even if you gave it an easy one like Jay Leno's chin).

There are several techniques that we will discuss later in this chapter that use "subpixel detection." However, there isn't enough information in a high resolution satellite image to extract that detail.

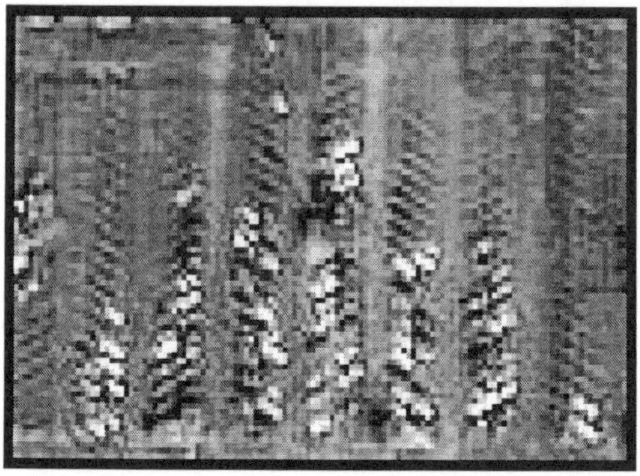

Figure 8.1

A parking lot as seen from space. One meter resolution. (Space Imaging)

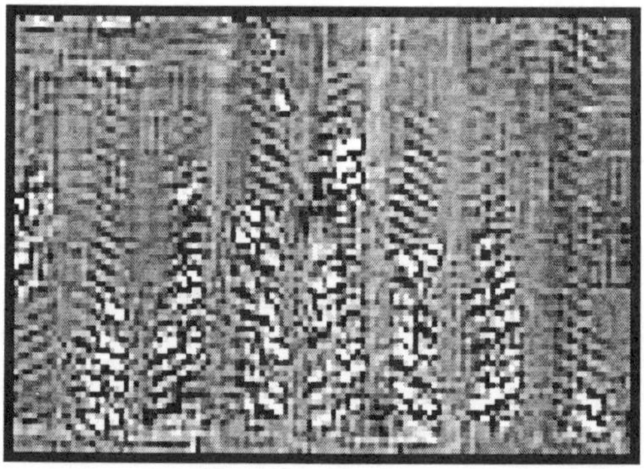

Figure 8.2

A sharpened image of Figure 8.1. It doesn't help as much as Hollywood thinks. (Space Imaging)

You can read a car license plate from space. Okay, who is putting their license plates on the roof of their car instead of the bumper?

Under the best of circumstances, with a good image and good software to enhance the image, you might be able to get the numbers of the plate (forget those mottos like "Land of Enchantment"). But there are a lot of problems that have to be considered. For instance, the best resolution images that might see that type of information are panchromatic. Since the modern vehicle license has a variety of colors on the background, the data would be confusing. To add to the confusion, remember that each state has several types of plates (regular, veterans, specialty causes, universities, public vehicles, etc.). It would be more of a challenge than you think.

A final proof that this is more legend than fact. Many jurisdictions have cameras that monitor traffic lights and send tickets to red light runners. Thanks to specialty films and spray-on products that can be put on a license plate, these cameras can't even read a license at 50 feet. Imagine the problems at 500 miles!

Satellites can see you in your house. Nope. True, an infrared sensor can see if a house is hot, but it can't detect someone in it. In fact, an infrared sensor can't detect someone standing behind a clear plastic sheet 50 feet away (a fact that Army and Marine snipers use to good effect).

If satellites were so good at seeing inside buildings, they would have had better luck finding the weapons that Iraq secretly built.

A satellite can follow you in your car as you travel. If that were true, maybe they could help me find my car in the mall parking lot.

As unique as you think your car looks, there are thousands of others that look exactly the same. When you consider that the satellite can only see the top of the car and items bigger than two to three inches in diameter, that number grows dramatically.

Now consider how extensive the American road system is and that within a couple of hours, your car can be on any road in about a 150 mile radius (or in underground parking for that matter). The chances that they were still following your car instead of someone else's are pretty slim.

Satellites can track anyone in the world. Despite all the resources used, they couldn't keep track of Saddam Hussein or Osama bin Laden. Enough said.

What Satellites Can Do

So far we have talked about what a satellite can't do. That doesn't mean that they are worthless. They have some pretty amazing capabilities that have an impact on your privacy. Here are some of them.

Detect minor changes anywhere in the world. If you add a deck to your house, you can be pretty sure that the local tax assessor will discover it if they use satellite imagery for assessing property taxes. Ironically, this development came out of the technology developed to monitor Soviet facilities during the Cold War.

Detect different vegetation. If a farmer changes the type of crops he is growing, if a lumber company cuts down some redwoods, if a drug dealer plants a new crop of coca in Columbia, or if you over water your lawn, satellite imagery can detect the change and highlight it.

Detect activity in and around buildings. Since satellites can detect heat, they can detect activity in a factory or a previously abandoned building now in use. The satellite, however, can't tell if the activity is harmless or something illegal.

See underground. Satellite-based radar can see a small distance underground depending on soil conditions. The dryer the soil the deeper the radar wave can penetrate. Also, thermal detectors can see the heat radiating through the soil or from air vents.

Detect metal objects nearly anytime. Metal objects are easy to find during the day, the night, under cloud cover, under trees, etc., because they are very good at reflecting radar waves.

Detect pollution. Thanks to multispectral imagery, satellites can detect many of the subtle changes that occur when pollution is released. Plant health, the chemical composition of the pollution, turbidity of the water, and smoke are all telltales of pollution.

Detect roads and trails. This may not seem like much, but it is important for the military, which may insert into a foreign country. Not only do they want to know what roads they can take to reach their target, they want to know what roads the enemy can use to counterattack.

What Satellites are Likely to Detect and How They Do It

Although satellites are a threat to privacy, we still have one advantage — there are more satellite images than there are trained satellite image interpreters. Consequently, no one is capable of monitoring everything that is going on — yet.

The key to tracking people and nations centers on a complex series of software developed by several computer companies. These technologies determine what can be easily detected and what remains hidden. Unfortunately, many of the papers that outline the calculations that go into this software are so technical they would make a rocket scientist go to sleep. Rather than force you to read technical papers like, "Orthogonal Subspace Projection with Local Background Estimation for Anomalous Subpixel Target Detection," I will give a brief, but understandable, explanation of the major computer software systems.

The most important software systems for automated satellite imagery processing rely on change detection. A computer looks at an older image and compares it to a new image. It then highlights any dramatic changes. If it doesn't see a change, it doesn't register anything.

This was originally developed during the Cold War. A satellite could look at several areas along the West German/East German border and automatically detect the movement and accumulation of men, machines and material that would signal a possible attack.

The change the computer looks for can be both spectral and spatial. For instance, if you build a deck on the back of your house, the satellite will detect a geometric shape that is attached to the back of your house and a spectral response that corresponds to whatever you made your deck from. The computer can then eliminate certain spectral responses like vegetation that would indicate that you have merely planted new grass. It can also eliminate certain irregular shapes.

Change detection software does have its weaknesses because it relies on finding a change, and categorizing the spatial and spectral information. If, for instance, someone put Astroturf in their back yard because its easier to maintain and then later built a deck covered with Astroturf, the satellite would have a hard time recognizing it because there is no dramatically different spatial or spectral data between the two sets of satellite imagery.

Conversely, dramatic changes from rains, earthquakes, tornadoes, or other natural disasters will overload the computer and interpreter. The change will be noted, but there will be so many other changes, the data from the image will be nearly worthless.

One way to make change detection better is to catalog the spectral responses for certain specific targets. In a military scenario, a CIA computer will have the exact spectral signature for Russian tanks, armored vehicles and aircraft. That way, if the imagery suddenly detects that specific mix of spectral data in a satellite image, it will flag it for a human analyst. On the home front, during a water shortage, the water police can look specifically for the spectral signature of well-watered grass so they can visit that homeowner.

This type of software also makes lower resolution multispectral imagery more useful in some circumstances. Even if the object you are searching for is smaller than the resolution of the image, you can find it, if the spectral signature is unique enough. If you are looking for a tank, when the smallest detectable object is 100 feet in diameter, you can look for the special spectral response of that tank's camouflage paint and see if it is found in any pixel. Depending on the strength of the spectral signature, you may have one or more tanks inside that one pixel.

Of course, paint can be changed, so computer specialists also developed what is called object extraction and pattern recognition, where computers can identify specific objects like aircraft and tanks. Computer software has been developed that allows a computer to recognize an aircraft and even identify it. This is a complex interpretation process because there are numerous types of aircraft and they look different even when viewed from a slightly different angle.

Of course, what software a specific government agency uses depends on what it is looking for. Zoning, taxing, and permitting agencies will look at change detection to see what changes have been made. A group monitoring a treaty will have specific software that will detect illegal equipment like certain aircraft or tanks, along with change detection software. Other intelligence groups may require software that shows lines of communication or highlights the massing of troops on a border.

Hiding from Satellite Imagery

Why hide from satellite imagery if you have nothing to hide? That's a good question. One reason is that as good as computers get, they will still make mistakes, and when government computers make mistakes, people get hurt. For instance, one of the biggest producers of automated satellite imagery interpretation software is SAIC, Inc. In addition to software for satellite interpretation, they were responsible for the U.S. government's Background Check System that has to okay every gun purchased through a firearms dealer. When one considers the criminals who got a gun with this system and those law abiding citizens who were denied a gun because of a computer mistake, it is easy to realize that the next mistake made by a computer may be made to you. Therefore, by knowing how one attracts attention from automated satellite imagery interpretation software and how to avoid detection, you know what you are doing that may unfairly raise a red flag.

The most important reason is the right to privacy that we expect in the United States. Hiding from government bureaucrats who want to control every part of your life is one way to keep that freedom. This is even more important in an era where the war on terrorism has begun to infringe on our expectations of privacy here in the U.S.

So, with that in mind, here are some general recommendations for avoiding unnecessary satellite intrusion.

Avoid attention and changes. As we noted earlier, change detection is the key to most automated satellite imagery processing software. If you can avoid dramatically changing the shape or spectral signature of the terrain or the changes are made slowly over a period of time, the chances of being detected are lessened. In the earlier instance, the homeowner put in an Astroturf covered deck after putting in an Astroturf lawn. Iraqis often didn't put fences around secret installations because they knew that installing the fences would trigger the change detection software and tell the interpreters that something special was there.

The spectral signature is just as important as the shape. If you plant a garden behind the house one year and then install a redwood deck the next year, the different spectral signature will give you away, even if the shape didn't. On the other hand, if you mulch the garden with wood chips and then cover the deck with a wood shingle roof, the spectral signatures will be closer.

The spectral signature isn't limited to manmade items. If, for instance, you decide to go out into the wilds and build an illegal cabin, change detection software can detect the change from living trees to dead stumps. Of course, not every acre of federal land is checked on a regular basis, so a modern day pioneer can avoid detection if the evidence of habitation is limited (small clearing, no smoke from fires, limited heat signature, etc.).

Stay under the resolution threshold. Although high resolution imagery is used in many places, low resolution imagery is still used to cover large areas. While a tax assessor for a city might use one meter resolution imagery to look at homes, a Department of Interior survey of forests in Alaska would be foolish to use that same imagery because of the cost and time to process. Besides, most of the information would be worthless. (Do you really need to check every square meter of a multimillion acre forest?)

Forestry experts still use the cheaper LANDSAT imagery (28 meter spatial resolution), which means that they can't detect anything smaller than about 100 feet in diameter. Consequently, any small-scale activity in that forest will go undetected if it doesn't effect the surrounding vegetation.

Of course, as was mentioned earlier, computers can look at spectral signatures and detect objects smaller than a pixel. And patterns can give you away. A star pattern will identify a Russian-style surface to air missile site even though you can't see the missiles. So don't assume that you are invisible when you are smaller than the satellite's resolution threshold.

Avoid manmade patterns. Nothing will grab the attention of an interpreter or computer software than a manmade pattern like a square, rectangle or triangle. Objects should follow natural lines and be covered in such a way to break up any straight lines. If possible, follow natural patterns like stream and river lines.

Make use of natural cover. Cliffs, caves, and dirt coverings will be less noticeable to a computer or interpreter.

Think multispectral. It isn't enough just to hide from satellites with a camouflage net purchased from the local army surplus. That may fool a high resolution panchromatic sensor, but it will stand out like a sore thumb with multispectral sensors, especially those in the near infrared bands. There are camouflage systems that are designed to hide from multispectral sensors, but they are much more expensive.

Since manmade materials don't normally have the spectral signature of natural materials, it is wise to use natural materials when at all possible. Although dying plants lose their infrared reflection and stand out, freshly cut leafy material will blend better with the natural surroundings. The old material can be spread around the forests and fields to lower the surrounding vegetation's infrared reflection, which will make your structure less noticeable.

Of course, since multispectral sensors have worse spatial resolution, you can still remain hidden if you remain under the resolution threshold. For instance, camouflage netting can hide from one-meter resolution panchromatic sensors, and the poor near infrared signature of the netting won't be a problem if it is considerably smaller than the 4-meter resolution of the near infrared sensor.

Find a confusing background. A piece of terrain that includes a random pattern of bare rock, dirt, grass, and a variety of trees will confuse the detection software more. If a tornado or forest fire has recently hit it, the confusion is even greater.

Cloud cover is better than no cover. Although radar satellites can see through clouds, cloud cover does limit the number of sensors that can view you. Nor do you need total cloud cover. A partially cloudy day is quite effective because either the clouds or shadows hide much of the terrain. These scenes are very hard for change detection software to work with because the scene is complicated, changing cloud patterns throw change detection software off, and the chances that a particular part of the image is clearly visible in two separate satellite scenes are lessened.

Radar will work through clouds and detect large structures, especially metal ones. You can lessen the chance that you will be found by eliminating or hiding reflecting surfaces that are strong radar reflectors. They can be covered with a thick layer of absorbing materials like logs or dirt.

You can also use natural strong radar reflectors like cliffs, faults, and fractures to hide metallic objects. The radar return from these structures will be so strong that it will wash out the reflection of the metallic materials.

Camouflage the surrounding area and use decoys. By using camouflage decoys over a large area, you confuse the interpreter and software about where you might be. It forces them to look at more targets, creates a bottleneck in interpretation, and makes it easier for you to slip through the cracks.

Avoid the times when satellites are overhead. There are web sites that continuously track satellite orbits and help you predict when they are expected overhead (see the Appendix). Software is available for free online that will help in tracking, along with data on specific satellites. Unfortunately, much of what was online has been taken off because of fears that terrorists were using the data to hide from spy satellites.

If predicting the satellite's orbit seems too hard, you can avoid many satellites by avoiding the time they are likely to be overhead. In order to get the best images, most commercial satellites fly overhead in the mid morning. Air temperatures are cooler in the morning and there is less distortion caused by hot air or pollution. The added benefit is that shadows aren't long enough to hide many objects, but long enough for accurately measuring heights. Therefore, activities done at other times of the day are less likely to be monitored by satellites.

Summary

Although the growing constellation of surveillance satellites poses a threat to our privacy, there are still ways we can keep our lives and activities secret. Satellites can't see everything and the computer software that looks at the imagery has weaknesses. You can use these facts to keep others from learning even more about you. However, as with other parts of our lives, the question you have to ask is how much inconvenience you are willing to endure to gain that additional degree of control over your life.

Chapter Nine
Satellite Imagery and Three-Dimensional Visualizations

It wasn't that long ago that turning a piece of satellite imagery into a three-dimensional visualization required expensive hardware and software. Consequently, it was only used for flight simulation and some architectural design.

Today, an average personal computer can handle the hardware and software requirements for spectacular 3-D uses of satellite imagery. And, in some cases, everything you need for the job is available online for free.

The Technology

There is a lot of computing that goes into changing a two-dimensional satellite image into a three-dimensional landscape that you can tilt and even "fly" through. That's why it wasn't until personal computers had the computing power to handle the millions of calculations in each scene that this became something that the common person could exploit.

Any virtual reality landscape, whether it is a computer game on your kids' Playstation or a flight simulation, is just a construction of polygons. These polygons, whether they are simple squares, pentagons, or complex unsymmetrical shapes built with straight lines, are sort of building blocks that allow the complex shapes and textures found in virtual reality. These polygons are then combined to create a complex three-dimensional image, much like many pentagons are stitched together to make the round soccer ball.

As the image is tilted, the computer makes millions of calculations of how these individual polygons would appear to the viewer. In the case of a moving image, like those found in a computer game, the computer has to constantly make calculations to show the changing perspective of the image. No wonder it takes a high power computer to make it happen.

To turn a flat satellite image into an exciting three-dimensional image, you need a satellite image and elevation data for the area you wish to work with. That, fortunately, isn't as hard as it appears. Several years ago, NASA launched the Space Shuttle Topography Mission. This mission provided the elevation of 80 percent of the world's land mass, with an accurate elevation every kilometer. Best of all, all this data is available on the Internet for free (see the Appendix). It is called a Digital Elevation Model (DEM) and it is critical for 3-D satellite imagery.

What the computer software will do is turn this DEM into a landscape of polygons. It will then drape the satellite image of the virtual terrain. Once this is done, you can rotate the picture so it looks like you are viewing the landscape from a mountain, a tall building, or a plane. Much of the commercial software will allow you to "tweak" the data so you can elevate a building that the DEM missed or add texture to a group of trees. You can even stitch satellite and digital imagery together to give greater resolution to some areas of the scene. If you are developing a flying simulation, you can change light and fog effects to make the scene look even more realistic. You can even add buildings or reconfigure the terrain.

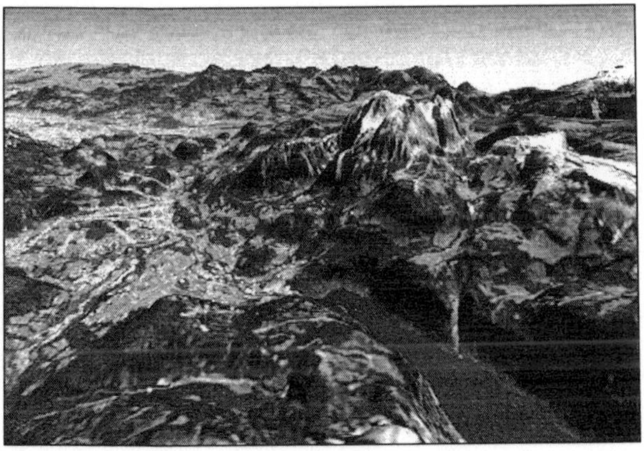

Figure 9.1

Salzburg, Austria as seen with three-dimensional imagery. (Tru-Flite)

Exploiting 3-D Imagery

The applications for 3-D satellite imagery are just as great as with the flat product. In fact, they are even greater because people relate better to three-dimensional imagery than overhead pictures. It also allows you to better communicate complex information to the average person. Here are some examples.

Presentations for public relations. Let's say you are a developer and you intend to build a single-family housing complex, if you can convince the local government and public that you aren't threatening the habitat of the rare 36-toed three-eyed newt. You know that the newt lives in low areas that are likely to be flooded during rains. You have accounted for that, but the engineering map of the development and the flat satellite image of the area doesn't show that you have spent considerable time and effort to protect every low lying part of the land.

With a three dimensional image of the area, you can show the skeptics that you are only building on high land and that all the low land has been set aside as a preserve for the newt. You win the public over with your concern for their precious critter and you win easy approval from the city board.

Nearly any presentation that you can make with flat imagery will look better with three-dimensional views. Not only does the different angle make it more understandable to the average person, you can get your point across easier. For instance, the low lying land in the housing development may only be five feet lower than the rest of the property, which would make it hard to show in the traditional 3-D image. However, you can change the ratio of the elevation so the variation in elevation is more dramatic and easier to show to a group of untrained viewers.

History. We have covered many of the historical applications of satellite imagery. Just imagine the uses of three-dimensional imagery (not only for yourself, but for sale to producers of historical television shows). One could use elevation data from a Civil War battlefield (one now covered up with development) and old satellite imagery to create a fly over of a now gone battlefield. With commercial software, you could add men and cannons to the scene to make a realistic view. You can develop a view like that seen by the generals and by adding fog, show how visibility was restricted during the days of black powder.

This technology isn't limited to satellite imagery. There are a lot of historical overhead images available from the National Archives that are just begging to be used. Imagine a realistic view of no man's land in World War One, a three-dimensional fly through of a Nazi death camp using aerial intelligence imagery, old western towns that are now gone, but were around in the 1930s and 1940s when the government photographed much of the U.S. Although the photographic images would have to be scanned, the opportunities are limitless.

Tourism sites on the Internet. Imagine how a tourism bureau could sell their region with a 3-D movie clip giving the visitor to their web site a fly over of the most interesting sites. Depending on the needs of the visitor's bureau, all the imagery may even be free online, which makes it a real high profit margin endeavor. And, since every visitor's bureau needs something like this, the customer base is nearly endless.

Military and law enforcement. Imagine knowing the high points and how the view from them would look long before you get there. That information would be critical for forward observers, artillery units, sniper teams, and communications units.

This imagery can even be taken to a higher level with high resolution satellite imagery that offers one meter or better resolution. A facility can be recreated in 3-D so a Special Forces unit can practice an operation before they insert. This requires more complex information than an Internet available DEM and more work by the operator who processes the information, but the potential for training exercises is immense.

Although the military applications stick out, many paramilitary police units need the same information for raids and hostage situations. In the case of a police force, the model could be developed long before a crisis and be readily available when the situation calls for it.

Local news. A news event is breaking, but the police and emergency vehicles have closed the area. What's worse, it's night and the news helicopter can't get a good view. Besides, the police helicopters are flying over the area and want you out. Rather than show the correspondent standing outside the closed area, use computer generated satellite imagery to give a view of the area while the correspondent narrates what is happening.

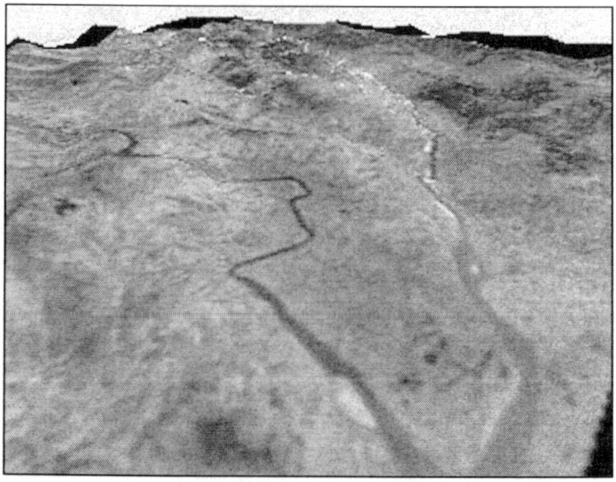

Figure 9.2

This image shows the approach to North Korea's nuclear facility at Yong Byon as it would be seen by an attack aircraft. It was put together using free satellite imagery and digital elevation models from government web sites. It took about one hour to produce. (The software was produced by Truflite.)

Summary

A little warning. Although there is free software and data available on the Internet (see the Appendix), you generally find you get what you paid for. The best software (with the tutorials necessary for making good use of the software), the most detailed DEMs, and the best imagery will cost something.

Although software for creating three-dimensional imagery runs the gamut from free to thousands of dollars, there are some sensible software packages out there that are understandable, flexible, and inexpensive. One such program is TruFlite, which was developed by an Austrian and costs a fraction of what other three-dimensional programs cost. The software is understandable and can even run on older computers that use Windows 95 (they even have a version that runs on Windows 3.x). Most expensive software requires the newer versions of Windows and more powerful PCs. You also have to pay for expensive updates. TruFlite offers free updates and actually has a person who will answer the phone and walk you through any problems you have. Free updates, low prices, and a person you can talk to when you have a problem. What more can you ask for?

Appendix I
Picking the Right System for Your Needs

Although satellite systems come and go, here is a guide of the different industries and applications and how well types of satellite imagery meet their needs. For the sake of this chart, high resolution means five meters or less, medium resolution means six to 30 meters, and low resolution means anything worse.

INDUSTRY	EXCELLENT	GOOD	POOR
Agriculture	HRMS, HRMS, HRHS, MRMS	LRMS	LRP, R
Archeology	HRP, HRMS, HRHS, MRMS, R	LRMS	LRP
Business Intelligence	HRP	HRMS	Others
Construction	HRMS, HRP, MRMS, MRP	HRHS	LRMS, LRP, R
Environmental	HRMS, MRMS, HRHS	HRP,MRP	R, LWP, LRMS

INDUSTRY	EXCELLENT	GOOD	POOR
Forestry	MRMS, HRHS, HRP, HRMS	LRMS, LRHS	R
Local Government	HRP, HRMS	MRP, MRMS	HRHS, LRMS, R
Insurance	HRP	HRMS, MRP	R, HRHS, LRHS, MRMS
Mapping	HRP, HRMS	MRP, MRMS	LRMS, LRP, R
Mineral Exploration	HRP, HRMS, R, MRMS, R, HRHS, LRHS	MRP, LRMS	LRP
National Defense	HRP, HRMS	MRMS, R	MRP, HRHS, LRMS
Real Estate	HRP	HRMS	MRP, R, HRHS
Transportation	HRP	HRMS, MRMS, MRP	LRP, LRME, HRHS, R
Utilities	HRP	HRMS, MRP, MRMS	LRP, LRMS, R

KEY

HRMS = High Resolution Multispectral (5 Meters or less)
HRP = High Resolution Panchromatic (5 Meters or less)
MRMS = Medium Resolution Multispectral (6 to 30 Meters)
MRP = Medium Resolution Panchromatic (6 to 30 Meters)
LRMS = Low Resolution Multispectral (31 Meters or greater)
LRP = Low Resolution Panchromatic (31 Meters or greater)
HRHS = High Resolution Hyper Spectral (30 Meters or less)
LRHS = Low Resolution Hyper Spectral (31 Meters or more)
R = Radar

Appendix II Sources for Satellite Imagery and Other Resources

The commercial satellite industry is dynamic and subject to change. Consequently, rather than mention names and addresses of satellite imagery companies in the text, I've put that information here in the Appendix, where it can be quickly referenced (and updated in future printings).

Sources for Satellite Imagery

These companies are the primary source for imagery in the United States. Their web sites also offer sample imagery and ways to find and buy their products.

Central Trading Systems
American agent for Russian imagery
15 Terra Mar Drive
Huntington, NY 11743
631-427-8390
www.sovinformsputnik.com

Digital Globe: QuickBird imagery
1900 Pike Road
Longmont, CO 80501
800-655-7929
303-682-3800
www.digitalglode.com

EROS Data Center: U.S. government satellite imagery
47914 252nd St.
Sioux Falls, SD 57198
800-252-4547
605-594-6151
edc.usgs.gov

Space Imaging: Ikonos, IRS, LANDSAT, and Radarsat imagery
12076 Grant St.
Thornton, CO 80241
800-232-9037
303-254-2000
www.spaceimaging.com

SPOT Image Corporation: SPOT imagery
14595 Avion Parkway
Suite 500
Chantilly, VA 20151
800-ASK-SPOT
703-715-3100
www.spot.com

Great Satellite Imagery Related Web sites

Doc Savage's Archaeology Downloads. This is a good site for anyone interested in the uses of satellite imagery in archaeology, especially in the Middle East. archaeology.asu.edu/jordan/nimaset

Federation of American Scientists. This is an organization that closely monitors weapons proliferation and uses satellite imagery to track military developments. Although they have a clearly liberal bent that effects some of their interpretation of events, this site is a must for people interested in satellite imagery and national security. www.fas.org

Gateway to Astronaut Photography of the Earth. All the photographs of earth taken by U.S. astronauts since 1961. Good search engine. eol.jsc.nasa,gov/sseop

U.S. Gazetteer. Need a latitude and longitude for someplace in the United States? This Census Bureau web site will give you the location just by punching in your zip code. www.census.gov/cgi-bin/gazetteer

GEO Community. This is a great web site that looks at all aspects of satellite imagery and its applications. www.geocomm.com

Global Security. John Pike, who is a recognized expert on satellite imagery and national security issues, has helped to put this site together. Another must for people interested in satellite imagery and national security. www.globalsecurity.org

GOES Project Science. The Geostationary Operational Environmental Satellite provides the cloud images we see on the weather. Although the resolution is low, the coverage is vast and it does give broad views of areas. rsd.gsfc.nasa.gov/goes

GEONET Names Server. If you want to find a place and all you have is a name, you can go to the National Imagery and Mapping Agency's name finder. Not a satellite imagery web site, but the information is often necessary to get the right imagery. Gnswww.nima.mil/geonames/GNS/index.jsp

Hobby Space. If you want to learn more about how to predict when satellites will be overhead, this is a good place to start. www.hobbyspace.com

Institute for Science and International Security. Another national security website that uses satellite imagery. Their major concern is nuclear proliferation. www.iss-online.org

NASA Image Exchange. A database of NASA imagery from the manned missions. The resolution isn't the best, but the coverage is great and the price is right. nix.nasa.gov

National Imagery and Mapping Agency. Satellite imagery of the whole world is available from this site. The data from the scenes can take a lot of time to download. Geoengine.nima.mil

National Oceanic and Atmospheric Administration. A good source for overhead imagery that looks at natural resources. www.photolib.noaa.gov

Oddens' Bookmarks. The Fascinating World of Maps and Mapping. This is a good place to start in your search for maps or imagery. Thousands of links to map and overhead imagery web sites. oddens.geog.uu.nl

Remote Sensing Tutorial. If the technical part of this book excited you and you want to learn more about the science of remote sensing, buy my book *Satellite Surveillance*. If that isn't enough, go here. NASA's Earth Observing System maintains this site. rst.gsfc.nasa.gov

Shuttle Radar Topography Mission. The most precise images of the world ever produced. The goal of this mission was to accurately map 80% of the earth's landmass with incredible detail. This the critical elevation data necessary for three dimensional images using satellite imagery and "fly throughs." www.jpl.nasa.gov/srtm also srtm.usgs.gov/

Silver Surfer. A fun site that directs you to several databases that allow you to "download the earth." Maybe a little to ambitious, but much of the information available can be used for three dimensional images and fly throughs.
www.geocities.com/TimesSquare/1658/

Satellite Products and Services

There a lot of products and services available for satellite imagery, but these are at the top of the list.

Perry Remote Sensing. Although primarily a geological services company, Sandy Perry has done everything from look for gold deposits to sunken treasure with satellite imagery.
Perry Remote Sensing
22 Sedgwick Drive
Englewood, CO 80110
303-761-7777
Sandy@rm.incc.net

TruFlite. Makers of three-dimensional fly-through software that can be used with satellite imagery.
TruFlite
21026 Primrose Lane
Mission Viejo, CA 92691
949-395-1811
Anthony@truflite.com

Brenton White. Consultant on 3-D imagery, maps, and digital elevation models. He can answer many of the questions you may have for little cost.
Brenton White
Brenton White Company LLC.
970-203-0300
www.brentonwhite.com

Appendix III America's First Spy Satellite Mission — 9009

It's been over 40 years since the first spy satellite was sent up by the United States. Unfortunately, many of the people who participated in those exciting first days have died and much of the history of the program is lost forever. Although the CIA has produced histories of the project like the unclassified "Corona: America's First Satellite Program" and several secret publications, they have a lot of holes in them.

In order to preserve some of that undocumented history, a previously unpublished, personal history of the first spy satellite mission has been included in this Appendix. Carroll Lucas, a photo interpreter who was involved in interpreting that first mission, wrote it.

This account has been reviewed by the CIA and approved for publication.

KEYHOLE — 1
MISSION 9009

On August 19, 1960, the weather over the target recovery area, approximately 250 miles south of Honolulu was almost ideal, with just a few scattered white puffy cumulus clouds interfering with total visibility. The crew on the C-119 aircraft strained to catch a glimpse of their descending target, and was fortunate enough to detect the multicolored parachute while it was still at a high altitude. Using a specially designed trapeze-like apparatus suspended out the cargo door of the aircraft, the crew attempted to fly over the descending capsule and snag the payload by its parachute as it passed by. The first two fly-by attempts failed, but on the third try, the procedure was successful in catching the parachute along with its payload. The deployed apparatus, with its captured capsule was reeled into the aircraft and flown to Hickham Air Force Base, Oahu, Hawaii, where the capsule was transferred onto another aircraft bound for the United States mainland.

Much publicity was given to this complicated operation, since it was the first successful air retrieval of a payload capsule deorbited from an orbiting satellite. The captain of the C-119 aircraft was given the Distinguished Flying Cross, and the airmen all received the Air Medal. What was not revealed to the public was the content of the payload, 20 pounds of 70 millimeter original negative photographic film, the product from a high resolution camera residing on the orbiting Discoverer XIV satellite.

Discoverer XIV was the first successful mission of the CORONA program, a series of polar orbiting satellites publicly established to conduct Life Sciences research on the effects of a space environment on living organisms. This remained the cover story for the CORONA series throughout its lifetime.

In actuality, the program was designed to place orbiting cameras in a near-polar orbit to provide intelligence on the military capabilities of the Soviet Union, and to maintain global surveillance as needs arose. After thirteen failures, the first payload was deorbited successfully, just three months after Gary Powers, flying a U-2 aircraft, was shot down over the Soviet Union, effectively terminating deep aircraft reconnaissance penetrations into that important target country.

As the capsule, with its unprocessed film still protected within, was on its way to a secret processing facility on the mainland, other activities were taking place to prepare for exploiting the film once it became available to intelligence community photo interpreters. The Central Intelligence Agency's Photographic Intelligence Center (PIC), housed in the Washington, DC, area began preparations for exploiting the Discoverer XIV film load as soon as they received notice that the capsule had been successfully retrieved. Based on its knowledge of exploiting large quantities of film, derived from experience with U-2 film products, and the many rehearsals conducted in preparation for Discoverer missions that were unsuccessful, PIC became the focal point for multiple pre-mission activities. Discoverer XIV now became officially Mission 9009, the first photographic mission from space.

At PIC, approximate geographic coordinates of the beginning and end of each of the eight successful 120 mile wide passes over the Soviet Union, were plotted on 1:250,000 scale World Aeronautical Charts (WACs). Target descriptions and readout requirements were distributed to the various sections and branches within PIC. Collateral photographic and non-photographic data was collated on all the known targets that fell within the bounds of the mission profile. These packets were then provided to the photo interpreters (PIs) for use during their analysis.

Paralleling these efforts, another group of interpreters were preparing to leave for the secret film processing facility to analyze the intelligence quality of the film products and to monitor the multiple reproductions that must be turned out. The procedures used by this "Tiger Team" had evolved from PIC's experience with the U-2 film products and were developed to better support the exploitation of the second-generation duplicate positive transparencies provided to each PI team. The Tiger Team was also available for on-site analysis during emergencies.

The Tiger Team arrived at the film processing facility several hours prior to the arrival of the capsule. Film titling information was reviewed along with film reproduction requirements. The decision was made to separate the mission into its various passes, and to break down the longer passes into two parts if deemed necessary. Each frame of the mission will have its own titling containing the mission number, pass, frame number and date. The film would then be rolled onto already available 70-millimeter film reels and stored in individual metal cans, ready for reproduction. The reproductions would be in the form of second generation positives, made from the original negative, and third generation duplicate negatives made from the positives.

When the capsule arrived, it was immediately placed on a specially made cart and rolled directly to the front end of the original negative film processor, where it was securely attached. Under the strictest of dark room conditions, the film was then reeled directly from the capsule film reel into the processor. Inspections were conducted during film drying and at the end of the processing cycle to assure optimum quality.

Once the original film had been processed in its entirety, it was then brought to the film breakdown room where it was separated into its eight passes and prepared for titling. It was during this time that the Tiger Team provided support to the film processing personnel responsible for processing, breaking-down, and reproducing the first of the film products. It was also the time when they conducted film analysis to determine its photo intelligence utility.

Being accustomed to viewing the large scale high resolution imagery from the U-2 aircraft cameras, the dense, very small scale, poorer resolution of Mission 9009 was an initial disappointment to the Tiger Team. The 24-inch focal length camera on Discoverer XIV, orbiting at altitudes ranging from 116 to 500 miles above the Earth produced scales from 1:300,000 to 1:1,320,000 at nadir. Since the 30-inch frame length depicted a monoscoptic panoramic view covering approximately 120 miles perpendicular to the orbit path, and the less than 70mm frame width covered an area of approximately 10 miles along the orbital path, analysis had to be conducted using magnifications of at least ten times.

As the original negative was slowly and manually reeled over a light table, the film processing personnel, working with the CIA Tiger Team, evaluated the quality of the film, noting all anomalies and determining the percentage of cloud cover on each frame.

It was immediately noticed that the first frame of each pass was fogged due to light leaks in the camera which affected the unexposed film while it was at rest for the close to 80 minutes between passes. Pressure marks from various camera rollers were also evident as dense bands perpendicular to the film edges. Both dendritic and corona static discharges had occurred, causing fogged areas scattered throughout the film length. There were no gaps in coverage between frames except when the cameras were turned off. An approximate five percent overlap occurred between sequential frames within a pass to assure continuous coverage.

Cloud cover was extensive, leaving only approximately 35% of the total area covered cloud free. Regardless, the total cloud free area was greater than the cloud free areas acquired by the total of all previous U-2 flights over the Soviet Union.

The quality of the processed imagery was deemed very good, with both photo density and contrast rated excellent, where not effected by the above mentioned camera anomalies. The ground resolution ranged from approximately 40 feet at nadir and near the perigee of the orbital path, to 75 feet off nadir and closer to the apogee of the orbit.

The type and amount of intelligence that could be interpreted from Mission 9009 photography was analyzed. Long linear objects such as major roads, railroads, and transmission line right-of-ways could easily be observed. Land/water boundaries, and ground scarring that occurs during construction activities could be delineated. Quarries and mines that could be detected by their tailings and large gravel pits could also be identified. Croplands, forest cover, deserts, and mountainous areas were obvious, and urban areas were easily differentiated. Within towns and cities, major road grids could be observed, whereas the narrower roads and alleys blended into the unresolved small buildings and structures that comprise urban centers. Large park areas and large buildings could be detected, and industrial sites could be differentiated from conventional urban areas. Individual vehicles on roads could not be detected.

Surfaced airfields were easily identified, whereas small, grass covered airfields could not be reliably detected. Large aircraft could be detected on surfaced taxiways, but small aircraft could not. Surface-to-air missiles (SAM) sites could not be reliably detected, nor could other defensive anti-aircraft (AA) armament that did not adhere to known deployment configurations. Large military vehicle parks could be observed as occupied or unoccupied, but individual vehicle counts or vehicle identification could not be accomplished.

Ports and harbors were easily identified. Piers, breakwaters, large ship servicing equipment such as gantry and overhead cranes, as well as open storage could be detected. Large moored ships could be identified by class, and smaller ships could be detected. Ships underway at sea could be detected by their wakes.

Known missile test sites could be identified, and major changes due to upgrading, expansion, or new programs could be detected. Major known missile impact areas could be monitored for additional craters and other changes.

Once the entire mission's original negative film load had been broken down into individual passes, titled, duplicate positive and duplicate negative reproduction had begun, the Tiger Team returned to its Washington, DC, facility. The film processing personnel produced multiple copies of the original negative, and sent them to a pre-established prioritized list of intelligence components, and photographic laboratories. This was all conducted following the Top Secret Talent Keyhole requirements for control and dissemination of its products.

The original negative and three duplicate positives were sent to PIC as the first priority. Since the film would arrive within hours, and detailed PI analysis would begin immediately thereafter, preparations to support the exploitation of the film products were quickly initiated. The Pass number and approximate camera-on, camera-off coordinates for each of the eight collected passes were plotted on acetate paper covering 1:250,000 scale World Aeronautical Charts. The approximate width of the passes were also plotted, using the above coordinates as the center point of the frames. These Mission Coverage Plots (MCPs) were given to the PI components, to support mission exploitation planning, and to a contingent of collateral specialists assigned to support the PI's by collecting information about the targets that would fall within the areas covered by the mission.

Once the film arrived at PIC, the MCPs were updated to include occasional frame numbers within each pass, as well as cloud cover information designated as a code from 1 to 4, with 1 indicating no clouds and 4 indicating complete cloud cover. These plots were given to PIC components and were also disseminated to other PI units in the military/intelligence community.

When the PIC photo interpreters received their Mission 9009 rolls of positive transparency film, they viewed the film, using a Richards Corporation-designed 940 light table, which had a 40 by 11 inch fluorescent light source, arm attachments to accept film rolls, and a Bausch and Lomb zoom microscope designed to move to any position on the illuminated surface of the table. They manually cranked the film across the surface of the light table, viewing each frame monoscopically, a small area at a time, through the microscope, and making notes on each observed target of interest. These analyses were conducted, using whatever collateral information they had on the target, to further describe and update pertinent target data that should be disseminated. Since much of Mission 9009 coverage included areas never before acquired on film, it was necessary to position each target on a map, using precise geographic coordinates, so that reference to it could be made on future missions without fear of re-identifying it as a different target.

A list of priority targets, compiled by the Committee on Overhead Reconnaissance (COMOR), played a large role in establishing the daily planning efforts of the PI contingents. These were plotted out on the MCPs of the mission, and special attention was given to answering the requirements.

At a designated time, late in the day, the written data recorded by each PI was collated into an Immediate Report (IR). A special cable report, designated an OAK, was then compiled and sent out for wide distribution within the intelligence community. This cable described in detail all the information gleaned from Mission 9009 photography that day. Six OAKs were compiled and disseminated from Mission 9009 photography. As each day's OAK was released, PIC produced briefing boards on the most important intelligence derived during that day. The boards contained maximum enlargements of the priority targets, which were annotated to emphasize important aspects of the target and titled with the mission number and date of the photography. These boards were then briefed to the United States Intelligence Board (USIB) by Mr. Art Lundahl, the Director of PIC, and copies of the boards distributed to a select list of military/intelligence organizations.

Among the various targets documented on such boards were numerous newly discovered military airfields, SA-2 missile sites, whole new industrial facilities not known to exist prior to Mission 9009 coverage, new urban areas, and coverage of missile related test areas. Briefing boards were also produced on all targets covered that were on the COMOR high priority list.

The port city of Petropavlosk was photographed as well as such other military crucial areas as Komsomolsk, Volgograd, the Sarova Nuclear Weapons Research and Development Center, the Kapustin Yar Missile Test Range, and the city of Moscow. Significant new intelligence was acquired concerning these areas, which was documented in the OAK reports, and on briefing boards, for distribution.

Throughout the mission, twenty operational surface-to-air missile (SAM) sites, as well as six sites under construction, were observed and identified by the unique deployment configuration (previously identified on U-2 film) as SAM-2 sites. However, due to the limited resolution of Mission 9009, and cloud cover, it could not be reliably stated that all sites in the observed area were detected.

Well over 30 new airfields were discovered, and extensive road and railroad construction was noted. Many large areas of ground scarring were observed, some of which could be identified as new mining, and timber harvesting, while others could be connected to the expansion of large military or industrial facilities. A large number of new military/industrial complexes were observed, whose function could not be identified. These areas were plotted on maps and their coordinates were documented in the OAKs along with the information that functions were unidentified (UI).

When the last OAK was completed, all six OAKs compiled during the exploitation of Mission 9009 were consolidated into index form and published as a Joint Mission Coverage Index (JMCI) for maximum distribution to security cleared military/intelligence organization. The JMCI documented the target name, country, a short description of the target, pass and frame numbers, a universal grid set of coordinates, the cloud cover, the WAC map upon which the target fell, a special PIC identifier number, and the geographic coordinates of the target center. A chart showing the track of the collected passes was superimposed over a small-scale map, and attached to the report.

Detailed analysis of selected targets began as soon as the OAKs were complete. All sources of information from previous photography to human observations were used to expand or verify what was observed on Mission 9009 film. An extensive report, identified as a Photographic Intelligence Report (PIR) was then compiled on each target, over the weeks following the receipt of the film. Many already existing PIRs were updated with supplements based on new data acquired on Mission 9009.

While the detailed analyses was being conducted, several engineers involved in the development of the camera, and procedures of processing the film, convened at PIC to view the film and to be provided information on its utility towards answering community intelligence requirements. Supported by PIC personnel, they carefully inspected a second-generation duplicate positive of the entire mission and recorded in detail all anomalies that occurred. As examples, the positions of roller pressure marks were precisely measured to determine the diameters of the rollers causing the problem, and the positions of all static traces were recorded to examine their correlation with sensor commands and mechanical functions. The average spatial resolution of the mission was estimated and compared to pre-orbital expectations.

Overall, the mission was declared a resounding success, the system anomalies appeared resolvable, and the intelligence gleaned from its one and a half million square nautical miles of coverage, produced more information than the totality of all past missions over denied territory. The engineers were then able to further tweak those camera systems awaiting launch, and to alter the designs for the next block of sensors.

The experiences derived from both the engineering and intelligence analyses of Mission 9009 produced procedures that were followed throughout the CORONA program, evolving as the systems matured and the quality of its produces increased. A 1968 quote from President Lyndon Johnson, a major benefactor of the CORONA program states, “We’ve spent $35 or $40 billion on the space program. And, if nothing else had come out of it except the knowledge we’ve gained from space photography, it would be worth ten times what the whole program has cost.”

Glossary

Band: A segment of the electromagnetic spectrum. The part of the spectrum from .4 to .7 microns is visible light.

Blue Band: .4 to .5 microns.

CCD, Charged Coupled Device: A solid-state sensor that detects visible and some near infrared light. This is the preferred sensing device in most satellite imaging systems. It can't be used in many multispectral and hyperspectral applications.

Change Detection: The most common type of automated satellite processing software. The computer looks at two images and detects the changes from the older image.

Contrast: The range of tones in an image. A high contrast image ranges from pure white to black. A low contrast image uses only a small range of grays.

CORONA: The first United States spy satellite program. The first mission was 9009.

DK-1: Russian one meter panchromatic imagery.

Electromagnetic Spectrum: The wavelength range of electromagnetic signals. Ranges from gamma rays, X-rays, ultraviolet, visible light, infrared, microwaves, radio.

Enhancement: Methods for improving a satellite image.

EOS: Earth Observing System satellites, which are the core of NASA's Mission to Planet Earth.

Green Band: .5 to .6 microns.

Hyperspectral: Capable of detecting dozens of different bands of light.

Ikonos: A commercial satellite system that offers one meter resolution panchromatic imagery and four meter resolution multispectral imagery. Given its 98 minute orbit and 423 mile altitude, it passes a given longitude at the same local time (10:30 am). It can acquire imagery of the same spot every three days.

IRS: The Indian Remote Sensing Satellite System. Offers 5 meter panchromatic, 20 meter multispectral, and 180 meter multispectral.

Keyhole (KH): The U.S. military reconnaissance satellite system.

KFA-1000: A Russian satellite system that provided photographic 5 meter color imagery. Often used for mapping.

KVR-1000: A Russian satellite system with a nominal 1 meter resolution. Often degraded in commercial applications for national security reasons.

LANDSAT: A U.S. satellite system that offers multispectral imagery. Unlike many systems that only offer the blue, green, red, and near infrared bands, LANDSAT also offers mid and far infrared bands. It is a favorite of many scientists because of the many bands offered, though resolution is poor.

Middle Infrared: 1.5 to 2.35 microns. It can detect moisture content.

MK-4: A Russian system with a 10 meter panchromatic resolution. The photographic product was often used for mapping.

Multispectral: Capable of detecting many different bands of light.

NIMA: National Imaging and Mapping Agency. The U.S. military and intelligence community's satellite imagery experts.

Near Infrared: .7 to .9 microns. Best for detecting vegetation and plant health.

Panchromatic: A black and white image that records light from many different bands (usually blue through near infrared).

Pixel: Means picture element. This is the smallest unit of information on a digitized image. The equivalent of one dot on a TV screen.

QuickBird: A commercial satellite system with a panchromatic resolution of about 61 cm and a multispectral resolution (blue, green, red, near infrared) of 2.44 meters.

RadarSat: A commercial synthetic aperture radar satellite launched in 1995.

Red Band: .6 to .7 microns.

Resolution: Although this can mean either spatial or spectral resolution, it usually refers to the smallest object that can be detected by a satellite.

SAR: Synthetic aperture radar.

Spatial: The characteristics of an image that refers to shape or size.

Spectral: The characteristics of an image that refer to the bands of light (blue, green, red, infrared, etc.).

SPOT: The French system of satellites. Owned by the French space agency CNES. The resolution for satellites 1 through 3 is 10 meters panchromatic and 20 meters multispectral (green, red, near infrared). Spot 4 offers 10 meter multispectral (blue, green, red, near infrared). The planned Spot 5 will have 2.5 meter resolution.

Thermal Infrared: Energy emitted by objects because of their temperature. 10.5 to 12.5 microns.

Visible Bands: .4 to .7 microns. This represents the light visible to the human eye (blue, green, and red).

WMD: Weapons of mass destruction.

YOU WILL ALSO WANT TO READ:

competitors. And they'd pay good money to find out the facts like: your legal and financial problems; who your customers are; any violations of government regulations or codes; what new products your company has in the works. If it seems your competitors are reading you like an open book, maybe you ought to do a little reading yourself. ***1991, 5½ x 8½, 132 pp, soft cover.* $16.95.**

☐ **58075 SATELLITE SURVEILLANCE, *by Harold Hough*.** Once the exclusive tool of governments, satellite technology is now available to anyone who knows how to use it. Using actual satellite photos, helpful diagrams and simple language, ***Satellite Surveillance*** explores the many commercial uses of satellite imagery, including: Land and water navigation; Zoning and land use planning; Fighting forest fires; Pollution control; Environmental monitoring; Habitat creation and protection; Oil and gold exploration; Organizing disaster relief; Building roads, bridges, and dams; Earthquake preparation; And much, much more. ***Satellite Surveillance*** gives you access to this powerful technology. ***1991, 5½ x 8½, 196 pp, illustrated, photographs, soft cover.* $21.95.**

☐ **91254 ÜBERHACKER II, More Ways to Break Into a Computer, *by Carolyn Meinel*.** Warning: If you don't like our quest for freedom in cyberspace, or if you have a broken funny bone, stay away from this book — it will just make you mad. Is this book a one-stop school for computer criminals or a solid piece of work on crime prevention? What it is, is the best hacking book ever written, now completely revised, expanded, and updated. "You won't learn how to be an überhacker just by reading this book. However, you will get keystroke by keystroke instructions on how to break into computers, keep the bad guys out, and learn the basics of how to discover new exploits on your own." ***2003, 8½ x 11, 370 pp, illustrated, indexed, CD-Rom included, soft cover.* $39.95.**

We offer the very finest in controversial and unusual books — A complete catalog is sent ***FREE*** *with every book order. If you would like to order the catalog separately, please see our ad on the last page of the book.*